A MANAGEMENT ODYSSEY

The Royal Dockyards, 1714–1914

J. M. Haas

UNIVERSITY
PRESS OF
AMERICA

Lanham • New York • London

Copyright © 1994 by
University Press of America,® Inc.
4720 Boston Way
Lanham, Maryland 20706

3 Henrietta Street
London WC2E 8LU England

Library of Congress Cataloging-in-Publication Data

Haas, J. M.
A management odyssey : the Royal Dockyards,
1714–1914 / J. M. Haas.
p.   cm.
Includes bibliographical references and index.
1. Navy-yards and naval stations—Great Britain—Management—
History.   2. Great Britain. Royal Navy—Facilities—History.
I. Title.
VA460.A1H33   1993   359.7'0941—dc20   94–2742 CIP

ISBN 0–8191–9461–1 (alk. paper)

For Mary Jo and Kevin

Contents

List of Tables

Abbreviations

Add. MSS.	Additional Manuscripts.
ADM	Admiralty Papers, Public Record Office.
BL	British Library.
CL	William Clements Library, University of Michigan.
H.C.	House of Commons.
IND	Digests, Public Record Office.
NMM	National Maritime Museum.
S.C.	Select Committee.

Preface

The making of this book has necessitated sifting countless documents, bound, boxed, or bundled, amongst the Admiralty Papers in the Public Record Office, a task which was often merely tedious rather than rewarding, and reading the reports, and underlying evidence, of more than two score commissions and committees of inquiry printed either in the Sessional Papers of the House of Commons or elsewhere. The financial support which Southern Illinois University at Edwardsville and the American Philosophical Society (through the Johnson Fund) provided in these endeavors was invaluable. I wish to acknowledge the assistance I received from many persons, too numerous to mention by name, at the Public Record Office (before its removal from Chancery Lane to Kew), the British Library, the National Maritime Museum at Greenwich, and the Ministry of Defence, Naval Historical Library. Dr. R.J.B. Knight brought an important letter of Vice Admiral William H. Henderson to my attention. My colleague Professor Norman Nordhauser helped me to surmount the problems a tyro encounters in using the indispensable word processor. I am especially indebted to Dean Samuel C. Pearson for his generous cooperation, to Amy Milgrim, Stacy Ratermann, and Penny Wilson, who prepared the manuscript for the press, and to Norma Jean Hartlieb, who supervised them. Any errors are entirely my own fault. For my wife's patience and encouragement I shall ever be grateful.

Chapter 1

Introduction

The Royal Dockyards were called into being nearly five hundred years ago to serve the Royal Navy. Both the navy and the dockyards manifested Britain's growing might to the world and, being very large themselves, represented a large, and ever-increasing, fixed capital. The replacement cost of the entire fleet in the first half of the eighteenth century has been estimated at £2.25 million (less than the cost of one dreadnought in 1913), or 4 per cent of national income.[1] Warships evolved very rapidly in size when iron and steel liberated them from the maximum size imposed by wood (reached soon after 1800), from 9210 tons in 1860 (HMS *Warrior*, the first iron warship) to 29,150 tons in 1913 (the dreadnought HMS *Queen Elizabeth*), or eight times the size of a wooden line of battle ship.[2] Moreover, the bigger a ship was, the more it cost per ton. Direct expenditure for building, repairs and maintenance in 1912-13 totalled £17,258,650 (an increase of 530 per cent in twenty-eight years), or 0.72 per cent of national income, of which the navy's own yards accounted for £8,384,852. The dockyards had to keep up with the growth of the fleet, the growing size of warships and developments in technology. Thus, to give just a few examples, rather more than £700,000 were spent on improvements between 1765 and 1785; the small yard at Sheerness was rebuilt and enlarged at a cost of £1 million at the end of the Napoleonic War; the steam factory and related facilities erected at Keyham (Devonport) in the early 1850s cost £1.225 million; £6 million in improvements were made at Keyham in 1896-1907.[3] It was naturally desirable that the dockyards, in view of the very considerable sums spent in them and on them, should be managed well and give value for money. How well were they managed, however? The question generated considerable contemporary interest and numerous official inquiries between the eighteenth century and 1914; thus it is of interest to us also.

The dockyards began life unassumingly as a drydock at Portsmouth in the reign of King Henry VII. Henry VIII, a more martial prince, was their real founder, however, as he was of the English navy also, thrice enlarging the yard at Portsmouth, establishing other yards at Woolwich and Deptford, and creating in London a department to administer all three.[4] Thus the dockyards presented already in the sixteenth century those organizational problems, associated with size and the inability of the owner, or owners, to superintend their business directly, which did not occur elsewhere in British industry until the Industrial Revolution and which, when they did occur, produced the "first appearance ever [of management], in its recognisably modern form".[5] Other yards followed: at Chatham in the reign of Queen Elizabeth I, Sheerness in 1665 (at the beginning of the second Dutch War), Plymouth in 1691 (during the War of the League of Augsburg), and Pembroke just after the Napoleonic War. The yard at Rosyth did not come into use until 1916. Portsmouth, Plymouth (later renamed Devonport) and Chatham were in that order destined to be the most important yards. Woolwich and Deptford declined in importance after 1689, owing to the westward shift of naval war and their cramped size and inconvenient location so far up the Thames, and were belatedly closed in 1869. Today there are but two yards, Devonport and Rosyth (both under commercial management), and soon there may be only one. There were a number of small overseas yards also, for repairs and maintenance. It was the home yards which the Admiralty had in mind, however, when it spoke of the dockyards.

Much of the navy was built in private yards (today all of it is). In wartime repairs and maintenance were necessarily the dockyards' dominant activities, and most building as a result had to be contracted out. Thus, one-third of the fleet (although not as yet the biggest ships) was built on contract in the eighteenth century.[6] The dockyards even in peacetime might not have sufficient building capacity; 25 of the 42 dreadnoughts laid down between 1905 and 1914 were built in private yards, a number of which now specialized in battleships. Dockyard-built ships probably cost more; however, it was widely held in naval and dockyard circles that the royal yards built to a higher standard, even though contracts were performed under continuous Admiralty oversight. The United States navy held the same opinion about its own ships but concluded that the qualitative differences were too small to justify the greater expense and so built very few vessels in the navy yards. Only very exceptionally, however, were repairs and maintenance performed in private yards. Such work generally cost less in the dockyards; moreover, the exact extent of a repair usually could not be ascertained beforehand, and the question of how much should be paid was highly complicated and open to wide disagreement.[7]

Shipbuilding was essentially an assembly industry and labor-intensive. Wages in the early nineteenth century (before iron and steel) accounted for 60-70 per cent of the price of a ship ready for sea. Labor-productivity was improved by the innovation over time of steam-powered lathes, drills, hammers, platebending, shearing, punching, drilling and riveting machines (increasingly driven from the late nineteenth century by hydraulic, pneumatic or electric power); even so, wages in the mid-twentieth century were 32-46 per cent of the cost of a ship's hull.[8]

Dockyard employment until the nineteenth century was driven by war. The number of dockyardmen, only 1185 in 1687, shot up during the Wars of the League of Augsburg and the Spanish Succession, reaching 6399 in 1711 and hovering around 8500 in the next three wars.[9] These were enormous numbers for the times. Total employment associated with the ironworks of Ambrose Crowley, probably the biggest private business in the second quarter of the eighteenth century, was only a little more than 1000.[10] Commercial shipyards, on the other hand, were very small; even as late as 1851, only 13 of the 317 yards in England and Wales had 100 or more workers, although several had over 1000.[11] The dockyard workforce at the same time numbered 10,000 (3000 less, though, than in 1814).[12]

The dockyards from the 1840s until 1914 were kept humming by the triumph of steam over sail, iron over wood, and steel over iron, the rapid evolution of design, armor and armament which rendered warships obsolete after they had been in service just a few years, and by the naval arms race. The workforce by 1914 had risen to 43,000, or 26 per cent of total employment in British shipbuilding.[13] There were now some very big private builders also, like Harland and Wolff who in 1913 employed between 15,000 and 16,000 in their yard and engine works at Belfast.[14] As late as 1935, however, only 31 British industrial firms had more than 10,000 workers, and only 10 had more than 20,000.[15]

The Royal Dockyards were not well managed, however. That was the conclusion reached by every official inquiry, and the opinion of most First Lords of the Admiralty, between the late eighteenth and the early twentieth century. Why they were not well managed, and what was done about it, and with what effect, are the questions which this book seeks to answer.

The dockyards presented to impartial observers an image of being inefficient, high-cost producers vitiated by the vices of slackness, indifference and indiscipline. Efforts to transform this inglorious condition commenced already in the late eighteenth century and intensified in the next century under the impetus of growing parliamentary and financial pressure to reduce costs and increase output per worker. Success depended on identifying and resolving fundamental problems of organizational system, information resources, and human resources, and doing so, moreover, without the incentives of direct competition and profit which are the motors of private business.

The dockyard organization was highly centralized. The yard officials were strictly limited to executing the instructions received daily from headquarters in London (the Navy Board until 1832 and the Board of Admiralty thereafter) and by a vast multitude of ever-proliferating standing orders; thus they were disinclined to assume any wider or individual responsibility. They were for all practical purposes unsupervised, however. There was no general manager to coordinate the construction, store and (after 1838) marine engineering departments; that was done informally by the heads themselves. Moreover, the very heavy correspondence occasioned by overcentralization, and the necessity of having to perform an oppressive number of ancillary duties which should have been delegated, distracted the master shipwrights from the active management of operations.

The heads of the London departments comprising the Navy Board were collectively responsible for the yards (as well as for most of the other civil business of the navy) and collectively decided everything at their general meetings. There was no managing director, however, an oversight which made the Board's proceedings liable to fall into confusion. Nor was there sufficient differentiation of functions in the department concerned with ship design and construction, materiel, and yard management. The abolition of the Navy Board and the assumption of its functions by the Board of Admiralty (hitherto its nominal superior) in 1832 created other problems, by denying the permanent officials responsible, in effect, for naval construction and the dockyards the authority to issue orders and conferring that authority on a superintending lord of the Admiralty instead, and by isolating the various London departments from each other.

Complete and accurate information is the lifeblood of management. The vast quantity of information available in the shape of sundry and extensive accounts and records was, however, mostly unreliable; nothing could be known with any certainty even about matters as elementary as the expenditure of labor and materials or inventories of stores. An estimate was required if a ship was to be repaired but not if it was to be built. There were neither financial nor cost controls. Top management had almost no personal contact with the yards.

Human resources were inefficiently managed. Naval constructors were little educated in the higher mathematics and sciences germane to shipbuilding; they were chiefly trained as craftsmen and maintained a narrow craft mentality. Jobbery was commonplace in the making of supervisory appointments. Large numbers of workmen were elderly, disabled, or simply lazy. Wages, and hence the intensity of labor, were low. Industrial relations were troubled, and the workmen to a remarkable degree dictated the limits of managerial authority.

The process of identifying and resolving these problems was exceedingly slow, so slow, indeed, that the image of the dockyards as the twentieth century approached was not very different than in the eighteenth century, despite all that had been done meanwhile to put management on a sound footing and annual visitations by the Board of Admiralty. A particular problem might be identified yet be understood only imperfectly, or the means of dealing with it might be beyond the knowledge of the times. Old ways and prejudices, mind-sets and bureaucratic inertia, reinforced by the antiquity of the dockyard organization, were big obstacles to change of any kind. Indeed, most reforms down to 1885 were essentially attempts to adapt the system of the eighteenth century to the very different conditions of the nineteenth century. A brilliant proposal for restructuring and energizing management was turned down in the 1860s not because the First Lord of the Admiralty disagreed but because it was too radical and would have put too many noses out of joint. The first superannuation scheme was introduced in the 1760s, yet the yards were still burdened with elderly men nearly a century later. A measure, adopted in the 1780s, for appointing supervisors by merit immediately became a parody of itself; another go at the problem in the 1840s at least was on the right track; twenty more years passed, though, before the merit system was placed firmly on the rails. The professionalization of naval construction, begun in 1811, was only completed in 1883. Efforts to establish financial and cost controls still had not succeeded in 1885; nor was it yet possible to say with any degree of certainty how much a particular ship cost to build. The problem of supervision still had not been solved in the 1890s, seventy years after it was first attacked, and idleness was still a serious problem in 1914. The need to decentralize and establish individual responsibility was first identified around 1800; it was not until 1905, however, that the deed was actually done.

Organization and management by 1914 had nonetheless been transformed and largely freed of the baggage of the eighteenth century (the subject of chapter 2) by a process which divides naturally into seven distinct periods defined respectively by the implementation of the recommendations of the Commission on Fees and the reports of the Commissions of Naval Enquiry and Naval Revision (chapter 3, 1793-1815), the reforms of Viscount Melville and Sir James Graham (chapter 4, 1815-34), the confidential report of R.S. Bromley and the work of the Committee of Dockyard Revision (chapter 5, 1834-54), the Committee on Dockyard Economy and the Royal Commission on the Control and Management of Naval Yards (chapter 6, 1854-68), the radical reorganization by H.C.E. Childers, its failure, and the aftermath (chapter 7, 1868-85), the Committees on Dockyard Administration and Expenditure (chapter 8, 1885-1900), and the Naval Establishments Enquiry Committee of 1905 (chapter 9, 1900-1914).

Chapter 2

The Eighteenth Century

The system of organization adopted for the Royal Dockyards, and much favored in Great Britain and Europe in the seventeenth and eighteenth centuries, was characterized by a high degree of centralization and by the separation and collectivization of authority and responsibility. There were three levels of management: the Lords Commissioners of the Admiralty (the Board of Admiralty), the Principal Officers and Commissioners of the Navy (the Navy Board), and the principal officers of each yard. The Board of Admiralty, a ministerial body of shifting composition, was nominally responsible for naval administration in its entirety but in respect of the dockyards and other civil branches mostly confines itself to making policy. The Navy Board, appointed by, and subordinate and responsible to, the Board of Admiralty, was entrusted with all but a small part of the civil administration (the rest was under the Victualling Board and the Sick and Hurt Board). Local dockyard management was in the collective hands of the principal officers, appointed by the Board of Admiralty but taking their orders from the Navy Board.

Each dockyard comprised five separate, but interrelated, departments; however, it is only the two which survived to the twentieth century — the construction and store departments — that are of interest to us. The master shipwright was the dockyard officer first in importance; his department — the one with which we are primarily concerned — was responsible for the construction, repair and general maintenance both of ships and physical plant and accounted for 85 per cent of the workforce.

Under the master shipwright, but themselves principal officers nonetheless, were two assistant master shipwrights at the three principal yards (Portsmouth, Plymouth and Chatham) and one assistant master shipwright at the three secondary yards (Deptford, Woolwich and Sheerness). The rest of yard operations were under the master(s) attendant (two at the principal yards). The master attendant (always a former ship's master) was the harbor master and thus responsible for launching, docking and careening ships; he was responsible also for the maintenance of the ships in ordinary (i.e. the reserve) and was in charge of riggers and sailmakers. The duties and background of the other principal officers were clerical. The storekeeper received, stored and issued all materials for the master shipwright's and master attendant's departments and for ships' carpenters and boatswains. Thus we must be concerned with his department also. The clerk of the cheque was the paymaster and acted as a control on the storekeeper. The clerk of the survey kept the accounts of carpenters' and boatswains' stores issued to, or received from, ships putting in for repairs or maintenance.[1]

The system of management worked like this. The principal dockyard officers provided the Navy Board with information and might make recommendations but had no latitude to take decisions themselves and were merely responsible for carrying out the Board's directives. Indeed, not so much as hap'orth of work might be undertaken without the express authorization of the Navy Board, which, in turn, was obliged, although largely as a matter of form, to clear everything with the Board of Admiralty. Thus there was at all levels a very heavy consumption of paper and ink. The Commissioners of the Navy never acted except in their collective capacity; all decisions were taken collectively, at their general meetings (held two or three times a week in peacetime but daily in wartime), and communicated in the same way. This cumbersome system of "management by correspondence", as it came to be called in the nineteenth century, often created confusion and led to delays in production, although it is possible that the master shipwrights often did not wait for authorization before undertaking repairs (the practice, as we shall see, in the nineteenth century). Even though the Navy Board corresponded collectively with the principal dockyard officers and addressed its letters to them collectively, the officers themselves corresponded individually and directly with their opposite numbers in London. The principal officers were all equal; there was no general manager, and no one in particular was in charge, although the master shipwright was clearly the most important officer. The principal officers gathered early every morning to read the day's letters from the Board, execute their instructions, settle the distribution of labor, coordinate their departments, and make such other arrangements as might be necessary. The "morning meeting" was crucial to the functioning of the yard and, although often neglected, remained a prominent fixture of local management. This was the pattern that dominated the management of the dockyards until the twentieth century.[2]

One other official there was, the highest ranking in the dockyard yet paradoxically the least important. This was the resident commissioner, always a naval captain and nominally a member of the Navy Board, even though his duties were entirely local and he never attended the Board's meetings. The office, which only dated from 1690, was created so as to provide a naval presence and more effective control over local management. For nearly a century Portsmouth, Plymouth and Chatham alone had a resident commissioner, because of their distance from London. Meanwhile, the two senior members of the Navy Board, the Comptroller and the Surveyor, visited Deptford and Woolwich, and the commissioner at Chatham visited Sheerness, either weekly or as often as possible,[3] an arrangement that was not very satisfactory and early on broke down. By the end of the century, however, the secondary yards also had their own resident commissioners, Deptford and Woolwich from the time of the War of American Independence[4] and Sheerness after 1793.

The duties of the dockyard commissioner were supervisory rather than managerial. The most important of these was to enforce the Navy Board's regulations, orders and instructions, which were for this purpose sent to him to be delivered to the principal officers; he could, however, issue no orders himself, except in an emergency; nor, even though he had authority over everyone in the yard, could he dismiss anyone; nevertheless, he alone could recommend dismissal to the Navy Board and administer fines, in both cases normally on the recommendation of the principal officers. He additionally provided liaison between the large naval element at the port (*viz.* the commander-in-chief and the commanders of ships) and the principal officers of the dockyard and reported to the Navy Board on the movements of ships at the port. The yard commissioner's instructions were unclear, his role was ambiguous, and at least one commissioner (at Portsmouth in 1749) had never seen his instructions. Yet the Navy Board ignored the order given by the Admiralty in 1749 to draw up new and clear instructions, an indication, surely, of the importance the Board attached to the yard commissioner. The office was in fact a snug harbor, a reward for services rendered by naval captains whose career at sea was, for whatever reason, at an end. Most yard commissioners were not inclined to assert themselves and intrude upon the domain of the principal officers, who in any case did not suffer displays of independence and could count on the support of the Navy Board when push came to shove.[5] The extent of the dockyard commissioner's authority was a subject of controversy during most of the nineteenth century; meanwhile, his authority was ostensibly enhanced and the title was changed to superintendent. Nevertheless, Rear Admiral Sir Robert Spencer Robinson, the Controller of the Navy, in 1867 delivered this well-considered verdict: the superintendent was "a mere channel of communication" and "responsible for nothing".[6]

The number and variety of workmen in the master shipwright's department necessitated a large staff of junior managers (the inferior officers) and supervisors all of whom, unlike the principal officers, were appointed by the Navy Board. Naval construction was supervised by the foremen of the yard (one each at the principal yards for building, repairs, and maintenance), while under these were the quartermen each of whom supervised and inspected the work of a gang of twenty shipwrights (including four apprentices). There was also a master caulker, boatswain of the yard (in charge of unskilled laborers), master mastmaker, master boatbuilder, and masters of housecarpenters, bricklayers, joiners and smiths all of whom were assisted by supervisors. Of the workmen themselves, the shipwrights were naturally the biggest group by far, accounting on average for about 38 per cent of the total; unskilled laborers and sawyers, the next biggest cohorts, accounted for about 15 per cent and 6 per cent respectively.[7]

The Navy Board, like the local dockyard organization, originated in the reign of Henry VIII. Naval administration was until then in the hands of the Lord Admiral, who was chiefly concerned with the military side, and an assistant who managed the civil side. The growth of the dockyards led, however, to the appointment of additional officials (the original Navy Board), and these exercised "almost total control over naval matters under the general (and often nominal) direction of the Lord Admiral," to whom they reported monthly. Samuel Pepys reorganized the Board and in 1662 drew up the instructions by which it was governed until its dissolution in 1832.[8]

The Board in the eighteenth century consisted if between seven and eleven coequal commissioners most of whom were in charge of various departments at the Navy Office. The Surveyor of the Navy (a master shipwright) was the commissioner responsible for ship design, dockyard operations, materiel and physical plant (including design and construction), functions which were eventually differentiated in the nineteenth century. That was quite a lot for one man to do; thus in wartime (and eventually in peacetime as well) a Joint-Surveyor was appointed. The Comptroller of the Navy (always a navy captain) was, however, the senior and most important member of the Board; he presided as unofficial chairman at Board meetings and was a member of Parliament also. When the Board of Admiralty in 1749 visited the dockyards for the first time ever, the Comptroller was the only commissioner they took along with them. Pepys intended that he should control naval expenditure and function rather like the later Accountant General of the Navy, but although his department continued to be responsible for domestic bills and accounts, foreign accounts, and seamen's wages, the rest of the business was early on transferred to three clerical commissioners — the Comptroller of the Treasurer's Accounts (for checking the Paymaster's accounts),[9] the Comptroller of the Storekeepers' Accounts and the Comptroller of the Victualling Accounts — with the Comptroller of the Navy exercising a kind of oversight. The remaining members of the Navy Board were the Clerk of the Acts (in charge of the Navy Office secretariat) and the extra commissioners (usually navy captains) — one in peacetime, three in wartime — who helped out wherever they were needed.[10]

The Commissioners of the Navy and the principal dockyard officers were prototypal civil servants; their offices were considered to be a species of property, and they themselves were never dismissed. The Comptroller of the Navy during the War of the Austrian Succession, Richard Haddock, was senile and incapable of performing his duties, yet not until 1749 did the Admiralty make so bold as to superannuate him, along with three principal, and two inferior, dockyard officers who were in the same case. Meanwhile, the ancient and domineering Surveyor, Sir Jacob Ackworth, who had been in office since 1715, presided at the Navy Board. The Board of Admiralty repeatedly crossed swords with the old man but dared not see him off; he died in office a few months after Haddock's superannuation.[11]

The Admiralty was still an office of state when Pepys, fresh from his labors at the Navy Office, became secretary to the Lord Admiral, the Duke of York, in 1672 and, in order to give his royal patron better control over the Navy, set in motion changes which transformed the

Admiralty into a department of government. Most of the business appropriated by the Admiralty concerned navy personnel, although the Navy Board was obliged to surrender the right to appoint the principal dockyard officers also. The Admiralty and the Navy Office continued to share the same building in Crutched Friars until the Admiralty in 1699 moved to a magnificent new building four-and-a-half miles away in Whitehall. This physical separation of the two offices was a deterrent to communication and cooperation which removal of the Navy Office from its cramped quarters in the City to commodious Somerset House in the Strand in 1786 scarcely alleviated (despite an excellent courier service).[12]

The office of the Lord Admiral was executed by commissioners at various times in the seventeenth century and continuously after 1708. The Lords Commissioners of the Admiralty, all of whom were politicians and so came and went with changes in government, consisted in the eighteenth century of a First Lord, with a seat in the Cabinet, and six junior lords. The First Lord, despite appearance and common opinion, possessed all the power formerly belonging to the Lord Admiral but with this difference — all orders and commissions had to be countersigned by two of the junior lords (a mere formality). At least one junior lord, and sometimes the First Lord himself, was a naval officer, while Sir Charles Wager (First Lord, 1733-42), Sir Hugh Palliser (First Naval Lord, 1775-79), and Sir Charles Middleton (First Naval Lord, 1795; First Lord of the Admiralty, 1805-06) were former Comptrollers; otherwise, however, the Board possessed no professional expertise. The Admiralty Secretary (in charge of the office) was, as a rule, the only official thoroughly familiar with the Board's business; indeed, all First Lords depended heavily upon his knowledge of naval administration. He read all incoming letters, decided which should be laid before the Board (only about one in five was), and answered all correspondence. Since there were only four Secretaries between 1694 and 1795, they provided a continuity of administration which otherwise was lacking. It was undoubtedly a considerable help that Philip Stevens, who was Secretary from 1763 to 1795, had previously been at the Navy Office for twelve years.[13]

The administration of the spending departments of government by boards reflected distrust of the judgment, and even the probity, of individuals acting alone in positions of public trust. The arrangement worked well enough in the case of a policy-making body like the Admiralty whose administrative business was handled by the secretary. The Navy Board's business was wholly administrative, however, and the necessity of conducting it at general meetings seriously diminished the Board's ability to manage the dockyards. At these meetings instructions for the dockyards were discussed and authorized, accounts and contracts were examined and approved, the Navy estimates were settled, and all correspondence, even on the most trifling matters, and by the late eighteenth century with more than two thousand correspondents, was conducted. The correspondence with the dockyards alone was staggering. This highly disparate mass was brought to the Board higgledy-piggledy by the respective Commissioners and dealt with in similar fashion, the Comptroller presiding for hours on end over the random shuffling of miscellaneous papers.

Middleton, preeminent among Comptrollers of the Navy, has provided us with a vivid, although self-serving, description of the Navy Board at work, and of the muddle into which it fell, during the War of American Independence.

> The whole business [was]...transacted at one table..., as matters came by chance before us. No means had been taken for dividing the business, nor any provisions made for despatch and economy in carrying it on. Some members were overloaded with business, while others came and went as best suited their own conveniency; and it fell of course to my share to bring things to some conclusion out of this undigested heap, before the day ended, be it right or wrong. The natural consequences were — hasty decisions, accounts passed with very little examination, and a perplexing variety of opinions on the most simple subjects.
>
> Imagine but for a moment, a very great number of different subjects before a board thus constituted, in which a great part of the commissioners had very little interest, but could, at any time, delay business by raising doubts and starting objections, even while occasion pressed and necessity called for immediate decision that matters might not be retarded, nor suffered to run irretrievably into arrear. [It is hardly surprising]...that things were hastily done, and...many important accounts left to the examination of clerks... by introducing indiscriminately all matters as they arise at the board, the correspondence [with the dockyards], on which the whole depends, is continually interrupted, business of the utmost importance is delayed, and [settled] with little examination.... In this way of managing business many material branches are of necessity entrusted to clerks. In the comptroller's branch it is unavoidable.[14]

Management cannot well rise above the quality of information available to it; the Navy Board, however, "had no check on the yard expenses, nor any good information respecting the state of the stores." The information the dockyards provided the Board in the shape of various periodical accounts and returns, as voluminous as they were numerous, was very low-grade; most were so ill-constructed, carelessly prepared, or inexact as to be of little or no practical use; some, indeed, were only compiled for their own sake or having been requested once before (often for reasons long since forgotten) continued to be compiled afterwards.[15]

The dockyards were unquestionably the Navy Board's preserve; the Admiralty as a rule exercised only nominal control, and relations between the two boards were tranquil as long as this was so. During the War of the Austrian Succession, however, an activist Admiralty under the Duke of Bedford, exasperated by the Navy Board's failure to correct what it perceived to be serious shortcomings at Portsmouth, presumed to bypass the Board. The Navy Board characteristically met this challenge to its authority head on and declared in very high language that power over the dockyards was vested in the Board alone by its constitution and had "continued uninterrupted to the present time".

> The Lord High Admiral...and the rest of their Lordships Predecessors tho' by their constitution invested with the Civil and Military power of the Navy have always consulted this Board with regard to the Civil Economy of it and there is no instance of any alteration having been made in its Rules or Establishment but by the Advice and Opinion of this Board who have always been looked upon and treated as Councils to their Lordships in such Matters.[16]

Nevertheless, the efforts of their lordships to establish control over the Navy Board continued, intermittently and in the long run unsuccessfully, until the Board was abolished and the Navy Office was absorbed by the Admiralty in 1832.

The Navy Board's control over the yards, in principle growing ever tighter, was nevertheless tenuous in practice. The duties of most of the principal officers were laid down in the first instance in their general instructions, issued in 1662 (at the same time as those of the Commissioners of the Navy). The assistant master shipwrights, inferior officers and supervisors, however, had no general instructions; these officials were "guided in the performance of their duties by occasional warrants and orders from the Navy Board," yet many of their duties were "from time to time altered and regulated according to the ideas of their own Resident Officers, without any reference to that Board," so much so, in fact, that "an officer who had served half his life in one dockyard, if removed to another would find himself nearly as much at a lose to know his precise duty, as if he had never been in the service." [17]

The standing orders with which the Navy Board from time to time augmented the general instructions were a paradigm of the system, perpetuated for most of the nineteenth century, whereby the dockyards were managed. It was through the medium of the standing orders that official duties and administrative procedures were innovated, revised or discontinued and that irregularities and abuses were corrected. The number of standing orders grew at an ever-accelerating rate to 200 by 1715, over 400 by 1750 and 600 by 1771, after which they headed for the stratosphere, reaching 1200 in 1786 and by then dealing with almost every aspect of local management. Yet the standing orders were observed or not according to the principal officers' own views, so these attempts to improve local management and tighten the Navy Board's control were not always very effective. The Board, moreover, almost never visited the yards to see if their orders were actually being observed, except on those rare occasions when ordered to do so by the First Lord. The principal officers naturally did not call attention to their own omissions or snitch on each other, while the yard commissioners mostly blended into the local scene. Woolwich yard in 1749 did not have most of the standing orders, and Plymouth fifty years later had none at all, lacunae which seem in no way to have troubled the officers of either yard. Indeed, the chaotic state of the standing orders, multiplying ever more rapidly, arranged not according to subject but simply in the order they were issued, and without so much as an index, deterred busy yard officials from either studying or consulting them. Some orders were contradictory, while others had been superseded but not canceled. All were to be read quarterly to the assembled officers, but even this order was not always complied with.[18]

The Navy Board knew that they did not exercise very effective control but threw up their hands in collective despair and, in an admission of helplessness, were reduced to displaying prominently around the yards notices "offering encouragement and Protection to such as shall discover any Irregularities committed either by the Officers or workmen". Most informers were anonymous, and although some seem to have regularly supplied information, it is unlikely that intelligence obtained in this haphazard manner can as a rule have been very fruitful.[19] Even the rounders and watchmen (replaced by the dockyard police in the next century) who patrolled the yards looking for petty offences were not very

effective.   The Admiralty, however, relied on its own sources of
information, usually naval officers who kept their eyes open when their
ships were at a dockyard.   In 1710 it got wind of egregious abuses at
Plymouth and strictly ordered the Comptroller and the Surveyor to go
down and root them out.   Again in 1767 it ordered the Navy Board to
correct "the Disorders and Irregularities in the Yards [which] have
gradually increased...[as a result of the] neglect and relaxation of that
Discipline which it is the duty of the Officers...to maintain."[20]
Whereupon, the Board, thus prompted, visited the yards.

This organizational system, far from being unique, closely resembled
that of similarly circumstanced enterprises in the private sector.
Ambrose Crowley's ironworks in Yorkshire, established in 1680, were
closely directed from the firm's headquarters in London and were
nominally managed by a council of department heads who assembled
early every Monday morning to read and execute the weekly instructions.
There was a similarly elaborate system of paper checks and controls, a
"constitution" consisting of laws and orders which Crowley and his
successors repeatedly revised well into the nineteenth century, reliance
on informers, and, just as in the case of the Navy Board also, "a constant
sense of despair about the efficiency of control".[21]

The Admiralty, not the Navy Board, was the engine of innovation and
reform in a century which until the last decades was not notable for
either.   The Navy Board, indeed, was opposed to "any Innovation", as
it said in 1752, when the Admiralty, wanting to increase productivity,
ordered the adoption of task work (a species of payment by results).   Nor
was this the first time the Admiralty proposed task work; it had done so
in 1694 also; on both occasions, however, its determination crumbled on
the rock of Navy Board opposition.   The gentlemen players were no
match for the professionals.   The Navy Board, like most bureaucracies,
had a life of its own and was unreceptive to change, a conservatism
which mirrored, and was reinforced by, that of the dockyard officers.
Still, the Board was up to its ears in administrative detail; anything that
made even more work was likely either to be resisted or, as in the case
of the Admiralty's orders in 1749 to draw up new instructions for the
yard commissioners and in 1764 and 1767 to codify the standing orders,
simply pushed aside.[22]

The Navy Board needed to be shaken up.   The Earl of Sandwich, a
man passionately concerned, and immensely knowledgeable, about naval
affairs, was the first to make the attempt.   Sandwich served his
apprenticeship as a junior lord of the Admiralty from 1744 to 1748 under

his friend the Duke of Bedford, whom he succeeded (1748-51). He was first Lord very briefly again in 1763 before becoming a Secretary of State (1763-65), and finally from 1771 to 1782, for the last four years virtually as war minister, such was "his special knowledge of the highly technical naval situation" after France entered the War of American Independence.[23]

Sandwich returned to the Admiralty in 1771 wise in experience and determined to make large changes in the dockyards, so as to strengthen British naval power.

> In the various Offices in which I have served....the Crown, I have always found that 'til I understood the business thoroughly myself,I was liable to imposition, and fearful of taking any thing upon myself....
>
> The Naval Department is so extensive and branches out into so many different Channels [that it cannot be administered successfully unless] the First Lord...understands it fundamentally....
>
> In every branch of Business the best and most knowing Men have their prejudices, which frequently obstruct the Public Good; if an ignorant person (however high he may be placed) attempts to remove these prejudices, he will soon find himself unequal to the Task, and will be obliged to submit because he has no foundation on which he can defend his own Opinion....[24]

These words, written in 1775, have an eerily modern ring. Indeed, former cabinet ministers have by no means been unknown to complain today about having been manipulated by the mandarins in their departments. Even in the world of big business, it is often not the chief executive officer but lower-echelon experts, through whom crucial information filters, who really decide important questions. Sandwich, however, was not so easily led.

Sandwich it was who initiated general dockyard visitations as an instrument of control.[25] The visitation of 1749 was only the second ever, the first having been conducted by the Navy Board at the behest of the Lord High Admiral, King James II, in 1686. The purpose of the 1749, and of all subsequent visitations, was

> to examine into & become acquainted with the Abilities & Conduct of the Officers, the Sufficiency of the Workmen, the Condition of the Ships and Magazines, together with what Works were carrying on, that such Reformations might be made as should be found needfull to prevent any unnecessary Expense...to see to it that the Officers paid due Obedience to their Instructions, that the several Rules and orders for the Government of

the Yards were punctually carried into Execution, that the Ships of the Royal Navy were kept in Condition for Service, and the Money granted [by Parliament] was frugally expended.[26]

Conditions varied from yard to yard but generally confirmed the Admiralty's view that local management was impermissibly loose.[27] The standing order that was consequently issued embodied fifteen new regulations and, significantly, reissued no less than twenty previous standing orders.[28] Yet there was no follow-up. The next visitation, by the Earl of Egmont, occurred in 1764, after the next war. It was precisely after a war, however, that the need to inspect the yards was most urgent. For as Middleton observed, "As well might a banqueting room be expected to be left in perfect order at the end of a large feast, as the dockyards arranged in method at the end of an intricate war."[31]

Very little changed as a result of either the 1764 visitation or, we may assume, that of 1749. Sir Edward Hawke,[32] alarmed by the "present height to which "Disorders and Irregularities in the yards" had gradually increased as a result of the officers' negligence, in 1767 ordered the Navy Board to visit the yards annually and report on what they found. So general and, indeed, vague were the reports (doubtless by design) that Hawke, who was not to be put off, on one occasion made the Board rewrite their report and another time personally interrogated them.[33]

Sandwich saw that annual visitations could be not just an important instrument of control but the means of nurturing cooperation between the two boards. Every year from 1771 to 1778, when war with France supervened, he visited the yards accompanied by one or two of the junior lords, the Comptroller, the Surveyor, and the Clerk of the Acts. The accounts, in his own hand, of the visitations of 1771 and 1773-75[34] are a remarkable and invaluable record, even if they were written for the edification of posterity. The visitations themselves, in late spring and early summer, included a review of the fleet at Spithead, among other things, and took about a fortnight, although Sandwich, combining business and pleasure, went in the Admiralty yacht and so was away from London for about six weeks. The Navy Board, however, travelled more expeditiously by land and returned to the capital whenever the First Lord was at sea for any length of time. The visitors, attended by the principal officers (and the commissioner, where there was one), inspected the workmen and yard officials, the storehouses and workshops, and some of the ships in ordinary, meanwhile asking questions, making notes and giving orders on matters which either required immediate action or could be dealt with summarily. A working dinner was held on board the yacht at the end of the day, and when Sandwich returned to London the two Boards jointly reviewed the memoranda of the visitation, and general orders were issued.[35]

Sandwich wrote after the visitation of 1775:

> [A] Man is never so much convinced of what is right, or wrong, as when he sees it with his own Eyes; and...when People in inferior departments know that the Principal will look into things for himself, and form his Opinion on the Merit, or demerit of those under him from his own Observations; they will be less inclined to mislead him, knowing that it is for their Interest to recommend themselves to his Favour, by showing their Diligence & Capacity in the execution of their several Offices.[36]

How effective were Sandwich's visitations in improving management? There can, unfortunately, be no very satisfactory answer, for although the records are clear enough about what needed, and was ordered, to be done, they do not, as a rule, tell whether the orders had any effect. A fairly marked improvement probably did occur in the short run. Sandwich drew a sharp contrast between the state of the yards as he found it in 1771 (after four visitations by the Navy Board and one by the Admiralty in the previous seven years) and in 1775, claiming, with permissible hyperbole, a "visible alteration for the better, by a change from the utmost irregularity and Confusion into perfect order and Method".[37] Robert Gregson, who participated in the visitations in his capacity of chief clerk to the Clerk of the Acts, told Lord North, however, that what Sandwich saw was only the appearance of order and method contrived for his benefit; "a sudden visitation," Gregson argued, "is the only way to detect abuses and irregularities if any prevail," as he himself was quite sure they did in abundance.[38] Gregson's views were highly colored, however; he was a sycophantic client of the Opposition leader Lord Shelburne and deeply embittered against Sandwich.[39] Yet Sandwich was himself under no illusions about the magnitude of the problem and the need for vigilance; each visitation turned up fresh problems and examples of backsliding; indeed, he almost seems to have despaired of conquering the inertia which held the dockyards in its grip.

Whatever good Sandwich's visitations accomplished was mostly undone in the long run. He himself predicted in 1775 that if his successors did not follow his example, "many abuses will creep in, and tho' their Abilities may be greater than mine, they will not know so well

as I have done, how to administer the proper remedies."[40]   His
successors, however, did not follow his example. Lord Howe visited the
yards in 1784-85 and the Navy Board in 1792,[41] but those were the only
other visitations until 1802.

The second Lord Melville, influenced by Sandwich's example,[42]
revived and so firmly established the practice of annual visitations during
his long tenure as First Lord (1812-27, 1828-30) that his successors
never deviated from it.[43] By the end of the nineteenth century, however,
the visitations had lost much of their former importance as a management
tool; the railway made it easy to visit any yard whenever necessary, and
senior Admiralty officials did so routinely; local management had been
professionalized, and an array of bureaucratic controls vastly more
effective than any known in the eighteenth century had been developed.
Indeed, the visitations were nearly discontinued after the turn of the
century but were simplified instead.[44]

The Admiralty and the Navy Board between 1771 and 1782 worked
together more closely and harmoniously than in any other period.
Sandwich was exceedingly fortunate to have inherited as Comptroller his
friend and fellow-reformer Sir Hugh Palliser, appointed in 1770 and
translated to the Board of Admiralty in 1775. The effectiveness of Sir
Maurice Suckling, Palliser's successor, was seriously impaired, however,
by ill health for some time before death overtook him in 1778. Sandwich
next appointed Sir Charles Middleton (later Lord Barham). Middleton,
a reformer of enormous energy and considerable ability, was
unquestionably the greatest Comptroller ever, although neither his parts
nor his accomplishments were quite up the image he himself cultivated.
He was self-regarding, censorious and ambitious, qualities which
endeared him to no one, even though he was highly respected by many,
not least by Sandwich.[45] His kinsman Henry Dundas (later first Viscount
Melville and First Lord of the Admiralty) described him thus:

> He has very great official talents and merits, but is a little difficult to act
> with from an anxiety, I had almost said an irritability of temper, and he
> requires to have a great deal of his own way of doing business in order to
> do it well.[46]

Sandwich nevertheless had a close working relationship with Middleton.
Sandwich's successors, Viscount Keppel (1782-83), Viscount Howe
(1783-88) and the Earl of Chatham (1788-94), seldom consulted him,
however, and more than once he threatened to resign. Chatham took him

at his word in 1790. Nor did Middleton get along well with Chatham's successor Earl Spencer, who brought him to the Admiralty as First Sea Lord early in 1795; he resigned before the end of the year; Spencer continued to consult him, however.[47]

Middleton established the Comptroller's dominance at the Navy Board (at the cost, however, of straining relations, often rather badly, with his colleagues, especially the Surveyor, on whom he encroached most), imposed a degree of order and discipline, and in 1789 secured the appointment of a secretary to organize the business. He saw everyone else as being the problem and himself as being the solution. With breathtaking audacity he proposed to Sandwich that the Comptroller should be made a member of the Board of Admiralty and be given supreme control over the civil administration and patronage (no rivalry with the First Lord intended). That, needless to say, was not on the cards.[48]

Middleton told George III that upon becoming Comptroller in 1778 he had found the dockyards in a very bad state and alone by "the most persevering application and the most decisive conduct" transformed them, a claim strikingly like the one made by Sandwich in 1775. He was in his management of the fleet and the dockyards alike assiduous, prudent and enlightened, and left both in good physical condition for the next war. But much of the transformation, insofar as the local management of the yards was concerned, only existed on paper. Still, he left an indelible mark, although not always for the better: under him the number of standing orders, the best index of centralization, simply exploded. On the other hand, he attacked the long-neglected task with which the Navy Board had been charged in 1764 and produced an enormous manuscript volume in which all the standing orders down to 1786 were "digested under a few simple heads" and fully indexed. The volume was never printed, however, and the project was "laid aside and wholly forgotten" when Middleton resigned in 1790. He also revised certain returns and discontinued others that were useless, thereby beginning the war on paperwork which was fought, with mixed results, ever after.[49]

People are an organization's most valuable asset; it is upon their intelligence, ability, vigor and commitment that, other things being equal, the organization's success depends. Yet the most persistent complaints against the dockyards were against the very people who ran and worked in them. Avoidance of managerial responsibility was pervasive; the officers were said to be "supine and negligent" of their duties. When cases of mismanagement, negligence, or abuse came to light, the officers endeavored "to screen, to palliate, and to find excuses". Work-discipline was weak, worker output low, idleness rampant. It was, indeed, a scene of "Sloth and inactivity" that unfolded when the Navy Board visited the yards in 1767, and Sandwich in 1775 accused the officers of "conniving at the Negligence and Indolence" of the workmen.[50]   First-line supervision — one quarterman for every gang of twenty shipwrights and apprentices — was more than adequate in principle but lax in practice, the quartermen spending almost as much time loafing as with their men. Shipwrights were to be found idling at the tap house (provided at every yard until the 1830s), loitering about the storehouses, or just wandering around the yard.[51]   How is this state of affairs to be explained?

The education, practical training and selection of managers and supervisors are important factors in creating an effective organization. The shipwright officers, like the shipwrights themselves, served a seven-year apprenticeship.   There were, however, two distinct classes of dockyard apprentices, one consisting of lads articled to officers and destined (if they remained in the service) to become officers themselves, and another consisting of boys articled either to supervisors or ordinary shipwrights and rising no higher themselves.   There were no qualifications, except a minimum age (fourteen) and height and any others a prospective master himself might impose.   Officers and supervisors were entitled to a certain number of apprentices according to their rank (e.g. a master shipwright was entitled to five).   Only persons who had been apprenticed to a master shipwright ever became master shipwrights themselves, so the range of selection was rather narrow, although no narrower than in the world of business where the owner-manager was usually succeeded by a son or relative.[52]   The career path of the superior class of apprentices led through several years as working shipwrights and more years as quartermen (a very few served a spell in the Surveyor's office); for some it led no further than an appointment as master calker, master mastmaker or master boatbuilder, yet about half must have become assistant master shipwrights and most of those master shipwrights.   Although this does not account for all of the superior class, we do not know what became of the rest.   Nor do we know how much weight, if any, was given intelligence and management skills in promoting officers; promotion appears, however, to have been according to seniority.   Promotion to positions below assistant master shipwright depended heavily on the recommendation of the master shipwright himself; at those levels it was the best-connected, although not necessarily the best-qualified, men who were chosen.[53]

"Jobbing and favoritism" were commonplace in the dockyards (where were they not in that age?), and commonplace they remained until the middle of the nineteenth century. Nothing, from the inferior officers' positions down to the more eligible assignments coveted by workmen, was exempt, and local recommendations might be overridden by pulling strings in London.[54] The problem was particularly acute at the supervisory level, for the dockyard service was "almost hereditary", and each yard was dominated by powerful "clanships" knit together by blood, marriage, politics or other bonds, descriptions which even though drawn from the late nineteenth century are hardly less true of the eighteenth century, when, in fact, the incubus of favoritism was much on the workmen's minds.[55] The best-qualified candidates were sometimes passed over, even for candidates who were manifestly ill-qualified. The effect on the efficiency of the dockyards must have been considerable, although not for that reason alone: the system was demoralizing, and workmen gave only grudging respect to the officer who had procured his promotion by jobbing and favoritism.[56] Middleton alone condemned the system. He accused Sandwich of making improper promotions[57] (although later comparing him favorably with his successors), claimed that the Navy Board alone (meaning himself) was competent to decide who should be promoted, and demanded that the Admiralty should give back the power of appointing the superior officers. Sandwich, however, thought that he himself was just as good a judge and had no intention of enlarging Middleton's power. Promotions to supervisory positions were the greater problem, for relative merit was at those low levels much harder to ascertain. Middleton's solution was to introduce a primitive merit system which required the superior officers collectively to evaluate the character, ability and performance of all supervisors and of workmen eligible for promotion. The reports were generally so undiscriminating and vague, however, as to be meaningless, and the system was a farce.[58]

Shipbuilding was not yet a science but still an "art and mystery",[59] not yet a profession, like medicine or the law, but still a craft, even if the Surveyor did customarily receive the accolade. Nor were the men on the career-path to the top well educated. The superior class of apprentices spent the first five years working in the shipwright gangs and learning the craft; in the last two years they put their tools aside and learned draftsmanship and how to lay off a ship (i.e. transfer the plans in full size on the floor of the mould loft); then, their apprenticeship over, several years as working shipwrights followed. To formal education in mathematics, the sciences, mechanics, or marine architecture, however, they were strangers. The prospect of years of hard manual labor and close association with rough working-class men and boys must have deterred many middle-class families from choosing an otherwise eligible career for their sons. Yet it was the prosperous middle class that provided the superior class of apprentices; the high premiums — 120-200 guineas (or more) in the case of a boy articled to a master shipwright, depending on the status of the yard and the master's own prospects — and maintenance fees would have been beyond the reach of families of lesser means. Such boys would have been the beneficiaries of a sound general education already. The master shipwrights and other superior officers appear to have taken some care in supervising the training and education of their apprentices (the lads' families must have expected that they would) but cannot, in view of their demanding official duties, have personally instructed them to any great extent. The most intelligent and enterprising lads probably read on their own such scientific and technical works as came to hand and acquired some knowledge of shipbuilding theory.[60] It may have been with the intention that they would use the time to obtain some formal education that they were allowed to absent themselves occasionally from the dockyard (a privilege which was abused). When Sir Joseph Allen (Joint-Surveyor, 1747-49; Surveyor, 1749-55) was assistant master shipwright at Portsmouth, he sent his apprentice son for instruction in writing, drawing, geometry and mathematics at the recently established Royal Naval Academy.[61]

Ill-educated though the dockyard officers were, they were probably not less educated, in the early eighteenth century, than men in comparable positions in private industry. That changed, however, once the Industrial Revolution got under way and academies sprang up which provided scientific and technical education of use to manufacturers among whom the level of general education was rising.[62] The French, being more enlightened in these matters, in 1741 founded the School of Naval Construction in Paris where their future constructors received a rigorous scientific and theoretical education and were shaped into a professional corps. French warships as a result were better designed, although perhaps not better built, than British warships. Were the French dockyards better managed, however? The only work that touches on them ends with 1762, too early for the School to have had much effect; until then, at least, the French yards appear to have suffered from the same problems as the British yards.[63]

A few British naval constructors by the end of the century were sensible of "the disadvantages arising from the want of education as we advance in our profession," and the Commission of Naval Revision in 1806 attributed "the absence of...order and method" in the dockyards to "the want of Education in the Officers".[64] The officers having been trained merely as craftsmen took a craftsman's narrow view of their duty and focused on the immediate object, which was simply to produce a well-constructed vessel (or to repair it properly). Theirs was, moreover, a highly traditional craft; it did not require, and did not nurture, well-disciplined, methodical minds inclined to analytical thought such as the challenging complexities of managing a dockyard demanded.

Labor was on average much less productive in the dockyards than in private yards. For this the master shipwrights themselves were largely to blame. Indeed, output per shipwright varied widely according to which master shipwright was responsible for the work. The evidence is two-fold and quite remarkable. Before the Surveyor drew up the task-work scheme that was adopted in 1775, each master shipwright submitted his own scheme; these schemes varied, often markedly, however, in assessing the number of man-hours required to build ships of the various classes and each of their sections.[65] The Admiralty in 1832 asked the Surveyor of the Navy, the master shipwrights at Plymouth, Portsmouth and Woolwich, and a London builder, William Fearnall,[66] to estimate how much it would cost for the labor to build the frigate *Vernon* in six months. Fearnall, unlike the master shipwrights, did not state the number of shipwrights he would employ; fortunately, however, enough is known about wages and working hours not only to reckon approximately how many men he would have used but, more important, the number of man-hours involved in each of the four estimates. The differentials between the quantity of labor reckoned by Fearnall and the master shipwrights, and among the master shipwrights themselves, are striking, as is shown in Table 1. The master shipwrights would have used respectively, 28.5, 59 and 96 per cent more man hours than Fearnall. He, however, had the advantage of paying a task wage (with an underlying day-rate of 5s. 5d.), whereas the dockyards had to make do with a relatively low day-wage. Still, it is likely that if the master shipwright at Portsmouth had been able to pay by results, he would have built the *Vernon* in the same number of man-hours as Fearnall, assuming a 20 per cent increase in output per man, which certainly would not have been unusual. Not so the other two yards, however, which brings us to

Table 1.   Estimates of wages for building HMS *Vernon*, 1832.[67]

Builder	Wages £	No. of shipwts.	Average wage	Average hours per day	Total man-hours[a]
Fearnall	6120	115[b]	6s. 9d½d.[c]	9.5	170430
Plymouth	6480	216	3s. 9d.[d]	6.5	219024
Portsmouth	8040	268	3s. 9d.[d]	6.5	271752
Woolwich	9900	330	3s. 9d.[d]	6.5	334620

[a] Number of shipwrights x hours per day x 6 days x 26 weeks.
[b] Price ÷ wages per man for 26 weeks.
[c] Fearnall paid a task wage.
[d] The dockyards paid a day wage.

two other important points — the responsibility of the master shipwright for the productivity of labor and for controlling costs. The differentials between the number of shipwrights in the Plymouth estimate on the one hand and the Portsmouth and Woolwich estimates on the other are 25 and 53 per cent respectively, figures which spell the difference between good management and poor management. Some master shipwrights simply did not concern themselves very much with the cost of the work entrusted to them. Sir Spencer Robinson was right when he said in the 1860s that dear work followed certain master shipwrights from one yard to another.[68]

Accounts of wages and materials were generally unreliable and were never audited; the officially reported cost of any ship or work must, indeed, be construed very broadly.[69] The Auditors of the Exchequer might take as long as twenty years before checking the Treasurer of the Navy's accounts and did no more than make sure the arithmetic was correct. Cost accounting was in any case "essentially a product of the nineteenth century".[70] Treasury and Parliamentary control, which was much later to give impetus to financial and cost control, was little exercised; when naval expenditure exceeded the Parliamentary grants, as it often did, even in peacetime, the deficit was covered, in effect, by borrowing.[71]

Internal mechanisms for controlling costs did not exist, nor was there any interest in developing such. Neither were mechanisms for planning very effectual. The cost of building a ship (unlike the cost of a repair) was not estimated in advance, nor was it to be until the late nineteenth century. Standardizing the design of each class of ships, as the French did, might have been useful; the Surveyor, however, only furnished the general lines of the ship; the master shipwright filled in the details and might, indeed, modify the general design according to his own ideas. The task work scheme introduced in 1775 might have been used to control costs, but there is no evidence that it was. Survey reports and the weekly yard progress reports might have been useful both for controlling costs and planning, if only they had been reliable. A survey report was an estimate of the extent, cost and duration of a repair. The weekly progress report gave certain vital information: how many shipwrights were employed on each ship, how far the work had advanced (expressed in sixty-fourths in the case of new ships) and the estimated date of completion. However, completion dates were often unrealistic (Middleton tried unsuccessfully to correct this), and the other information was often equally unreliable, whether through carelessness or by design. The master shipwrights were often guilty of what Middleton called "unbridled" extravagance. They got away with it, though, by falsifying the progress reports and by charging part of the labor and materials for new ships and pet projects to vessels that were either being repaired or refitted, where the cost could easily be concealed without any chance of being detected.[72]

Private yards, since they confined themselves to building, could build a third-rate line-of-battleship in two-and-a-half years, whereas in the dockyards building and major repairs in wartime had to be sacrificed to smaller repairs and maintenance; thus it took much longer to build (thirteen years in one highly unusual case) and up to four years for a major repair. However, a certain amount of labor was routinely misapplied as the result of practices which neither the Navy Board nor the Admiralty saw fit to stop. Ships' captains might take advantage of their rank and insist upon extra work they happened to fancy (a practice which persisted in the nineteenth century), and large numbers of shipwrights were sent in wartime to make repairs to the ships at Spithead or the Nore or in Plymouth Sound which the ships' companies should have made themselves.[73]

Repairs and maintenance presented intractable problems for planning. Surveys were, as a rule, imprecise predictors, since additional work often cropped up after a vessel was taken in hand. Indeed, forecasting costs and outputs is still problematical today when the dockyards have come under commercial management.[74]   A ship proposed for repair was surveyed by boring into the timbers as it lay in the water; this was usually an inconclusive (but still the most practicable) method, and a much more extensive repair might often be indicated after the ship was docked. Decommissioned ships which had been reported as only needing a small repair were often subsequently found to be unfit for service; by then, however, they had been stripped too far to be floated out and had to be broken up in the drydock, thereby robbing the yard of the productive use of both dock and shipwrights for about a month. Ships were routinely refitted every three years in peacetime (less often in wartime); however, the innovation during the War of American Independence of sheathing their bottoms with copper made it possible to lengthen the interval considerably and thus was a boon to the yards. Three or four months usually elapsed in wartime between the time a ship of the line arrived in harbor and the time it left of which less than half was actually spent in drydock. The eastern yards (with the exception of Sheerness) were so inaccessible that small repairs and maintenance were largely thrown on the two western yards. These, however, were not provided with enough docks, and in wartime, as a result, ships often waited chock-a-block in the harbor. The docking of one ship seems to have been coordinated closely with the undocking of another; nevertheless, even the deepest docks were not deep enough; ships of the line could only be taken in or out on fortnightly spring tides and so might be held captive for some days after the work was finished.[75]

Low output per shipwright relative to commercial yards was owing also to differences in the composition of the two cohorts. The dockyards were lumbered with a large number of inefficient shipwrights. Henry Peake, the master shipwright at Deptford (and soon to be Surveyor of the Navy), reckoned in 1805 that 20 percent were superior, and 38 percent good, workmen; the remaining 47 percent were either old or lazy and a drag on the others. Peake's figures are validated by the fact that 41 percent of the shipwrights were excluded from task work in 1775 (the other 59 percent insisted upon it) on the grounds that they were either old or lazy.[76] The strength of the gangs was equalized by means of the annual "shoaling" (a practice which continued to be a fixture until the present century):[77]   the yard's shipwrights were assembled and the quartermen took turns picking a man until all the gangs were up to strength.[78] The practice in the private yards on the Thames, on the other hand, was to let the men of each gang reject any prospective workmate (or expel a workmate) who was not as strong and hard-working as themselves. For they were paid not by the day, as the dockyardmen were, but by the task, and their wages (and the builder's profit) reflected how hard they worked; the builders, indeed, would not hire men who were old and worn out. Thus the third category of shipwrights, so prominent in the dockyards, was simply unknown in private yards.[79] How did the dockyards come to be saddled with such men?

There was as yet no Establishment of workmen entitled to permanent employment and superannuation; nevertheless, dockyard-trained artificers were in practice employed for life anyway, and men hired from outside the dockyards might be too if they acquired sufficient seniority to survive reductions in force. The men themselves insisted upon this, and the Navy Board and yard officers saw it as the best way of attracting and retaining a loyal and disciplined core of workers, as well as sparing the local parishes the expense of having to provide poor relief. Worn-out shipwrights were paid the same as the able-bodied but were given light work appropriate to their condition, e.g. in the boathouse or masthouse. Shipwright's work required great physical strength, however, and the yards were forbidden to hire men over the age of either thirty-five or forty (depending on how much they were needed), yet men older than that were hired anyway, especially in wartime. At all times, however, older or otherwise unsuitable men got in, either by pulling strings or by the back door (i.e. without proper recommendation).[80]

Until the later eighteenth century, most workmen hired in wartime were discharged afterwards, even though the master shipwrights did so reluctantly. The visitation of 1749 turned up large numbers of workmen who had been hired during the late war (many of them improperly) and should have been discharged when it was over. The yards were subsequently obliged to let so many men go that the Navy Board complained (unavailingly) that not enough were left for essential work. After the Seven Years War, however, "none but indifferent hands were discharged, and but few of them, when compared with former reductions". Even so, the Navy Board in 1764 proposed to hire another 695, including 350 shipwrights. After the War of American Independence the workforce was rapidly run down by attrition (to a size even smaller than in the early 1750s), and only apprentices coming out of their time were hired, so as to bring on, in Middleton's words, "a succession of stout able men equal to any exertion."[81]

Superannuation, introduced by Egmont in 1764, was a humane way, acceptable to most able-bodied men, of ridding the yards of elderly or incapacitated workers. It was limited at first, however, to just two per cent annually of quartermen, shipwrights and caulkers. Superannuation might occur either after thirty years of unbroken service (at two-thirds pay) or at any time in cases of permanent disability (the pension being pro-rated). The first group to be superannuated, consisting of 105 men between the ages of thirty-five and eighty-five, tells a tale: 38 were in their seventies and 14 in their eighties. Simon Coward, an eighty-five-year-old Plymouth shipwright with sixty-four years service, was described as being "very Lame in both legs, Eye Sight impaired, quite enfeebled"; William Newman, a man of thirty-five with nine years service, had recently fallen "into the Stern of the double Dock [at Plymouth and] has ever since been disordered in his senses, [and] is now...deprived of Reason". Egmont's scheme only scratched the surface of the problem, however, and Sandwich in 1771 extended superannuation to all workmen and increased the rate to 2.5 per cent. Of the 118 men pensioned that November, the oldest was ninety-two, the youngest a man of twenty-eight, with only three years service, who had been blinded by "a great Cold". Indeed, the numbers superannuated rose markedly after the visitations of 1792 (to 314) and 1802 (to 905). Thus was the problem of aged and infirm men alleviated, even though it was by no means eliminated.[82]

The standard working day, defined for much of the year by the number of daylight hours and therefore seasonably variable, was similar in both the dockyards and private yards. For example, dockyardmen from March to September worked from 6 am to 6 pm, with time for breakfast and two hours for dinner (half-an-hour more than private yards allowed), for which they might leave the yard. In December and January they worked from 7 am to dusk, with one hour for dinner (taken in the yard).[83] In the dockyards (unlike in private yards), the workmen were mustered in a body (in military fashion), and each one's name was called, two or three times a day — in the morning, after dinner (if they were permitted to leave the yard), and in the evening — a process which might consume up to one-and-a-half hours. Private yards had no need to muster their shipwrights; each gang was the responsibility of its leading man, a private contractor through whom the men were paid.[84]

The dockyards' time-wasting system of taking attendance survived in modified form (mustering was eventually by divisions) until 1849, when it was replaced by a forerunner of the system used in industry today. The workmen entered and left the yard in continuous streams; each upon entering took from its hook on a board a metal ticket stamped with his number and deposited it in a box; the ticket was returned to him just before quitting time, and he replaced it on the board upon leaving (the dinner hour had been shortened, and the men were no longer allowed to leave the yard). Characteristically, however, the new system still had not been implemented everywhere a decade later.[85]

When the Admiralty in 1752 complained about how little work the shipwrights did, the Navy Board replied with an air of resignation that they did as much as could be expected considering how much they were paid — 2s. 1d. for a standard working day, a rate which had been in effect since 1690 and which there was no disposition to increase.[86] That was about as much as was paid at the West Country outports which provided Portsmouth and Plymouth yards with so many shipwrights. It was not with the outports, however, that the Navy Board was comparing wages, but with the London district. It was with London that the eastern yards (accounting as late as 1774 for nearly half the workforce)[87] competed for labor. There the day-rate from around 1740 to 1794 was 3s. 6d.; that was more than the Admiralty paid quartermen (2s. 6d.) or foremen of the yard (3s.). The master shipwrights at the other end of the scale were paid very well, their annual income from salary, fees and emoluments being in the range of £341 (Sheerness) to £508 (Chatham) and putting them at the lower end of salaries paid top managers and managing partners in the biggest firms (£300-£1000) and well above the salaries of typical managers (£50-£100).[88]

The day-rate was forty per cent less in the dockyards than in the London shipyards; however, the London shipwrights were paid not by the day but by the task and earned 17-100 per cent more than the underlying day-rate. Task work (called building for a price also) was a form of collective piecework in which the builder and the leading man of the shipwright squad negotiated a labor-price; the men were paid the day-rate as the work progressed and the balance of the price upon completion. In wartime, when demand was greatest, the shipwrights earned up to 7s. per day;[89] in peacetime they probably earned for the most part 4s. 1d.-4s. 7d. per day,[90] and occasionally, when order books were full, 5s. 6d.[91]

The Admiralty never contemplated raising the dockyard day rates, yet wages obviously had to be increased in proportion to those in London if the dockyards were to function with any degree of regularity and efficiency. In fact, a differential of about forty per cent for shipwrights was generally maintained by harnessing overtime (or "extra", as it was called) and using it to increase the pay of non-salaried supervisors also (e.g. foremen of the yard and quartermen). Overtime rates were double the time-rate per hour: 7½d. for a "tide" of one and-a-half hours and 2s. 1d. for a "double tide" of five hours. In peacetime, a single tide might be worked (during dinner time) for up to seven months a year; in wartime two tides were usually worked (the working day starting an hour earlier and ending half-an-hour later) and sometimes "nights" also (keeping the men in the yard from 5 am to 8.30 pm and making it necessary, except in summer, to work by candlelight). The ostensible purpose of overtime, in the words of the warrants authorizing it, was "to accelerate the building and repair of His Majesty's Ships". The Navy Board, extolling the advantages to a skeptical Admiralty in 1765, declared that "the works are advanc'd, in proportion to the Extra that is wrought", and the shipwrights themselves implied as much in their petitions for overtime. This was only true, however, in the sense that otherwise the men would have performed less work than usual, because only exceptionally did they perform more; overtime, in other words, was a subterfuge for doing what in the London shipyards was done openly by means of sliding wages. Richard Hughes, the energetic commissioner at Portsmouth at the time of the War of the Austrian Succession, opposed overtime because the shipwrights either could not or did not perform any additional work. Nor was Sandwich under any illusions about overtime; it was for this reason (among others) that he introduced task work in 1775, whereupon the shipwrights (then working two tides extra because of the insurgency in North America) struck, declaring that "in complying with the Task we are committing progressive suicide on our bodies." Yet they never complained about overtime being suicidal; on the contrary, they were avid for it. Overtime was an inducement for good shipwrights to enter the dockyards, but not because they were normally required to perform additional work, as the following Navy Board letter makes clear.

> It is worthy of notice, how greatly this encouragement of One Tyde Extra; for five, or seven months in the year, is a means of inducing the choicest Shipwrights, &ca. (whose pay is so very small) to give preference to the Kings yards; and also exciting them to a diligent exertion of their Duty.

Shipwright's work was exceedingly strenuous, and no one, not even the youngest and strongest, could have kept at it for thirteen or fifteen hours a day, except for very short periods; even eleven-and-a-half hours a day would have been counterproductive after awhile. The effect on attitudes to work of stretching ten hours of labor out to thirteen or fifteen hours can only have been pernicious.[92]

The apprentice system was another important way of augmenting wages and salaries. Everyone from the quartermen on up was entitled, depending on rank, to one or more apprentices, leaving less than twenty per cent of the apprentices to be articled to meritorious artificers. The privilege was extremely valuable, and much coveted by the workmen, since the master received, besides a premium, the largest part of his apprentice's wages.[93]

Little as their wages were, the dockyard shipwrights were probably better off in the long run than many of their brethren in the London yards. Shipbuilding was (and was to remain) a highly cyclical industry, and it was because of this, and the high cost of living in the metropolis, that the London shipwrights were paid so much. The younger, stronger gangs in slack times got all the work and many men were left destitute, for such a proud breed were shipwrights that only when their circumstances became truly desperate did they turn to alternative employment. Even when the men were in work, they might be interrupted, sometimes for days, by inclement weather, and time might be lost because of sickness or injury, for the work was hazardous. For old and worn-out men there was no work whatsoever, even in the best of times.[94] A unique system of social security, such as no friendly society could have provided, insulated the dockyard workforce from these vicissitudes. It was for this that wages were sacrificed. Lifetime employment or, after 1764, a pension was not the only boon, however.[95] Work of some kind was unfailingly provided every day (indoors if necessary), and after 1759 sheds might be erected whereby work on a ship could continue on wet days. Men injured on the job received free medical attendance (each yard had a surgeon) and were paid in full for the first six weeks they were off work and at a reduced rate thereafter until they recovered. Nor did they lose their jobs when serious illness kept them from working for an extended period, although this privilege was so abused that a time limit was imposed in 1774.[96] Nevertheless, none of this appealed to able London shipwrights or even, indeed, to many dockyard-trained men. William Shrubsole, the master mastmaker at Sheerness from 1773 until his death in 1797, wrote in 1770:

> It is very rarely seen, that a good hand leaves the [London] merchants for the King's service, unless he has some sinister views, from old Age or interest. But it is very common for the best hands to leave his Majesty's

yards and be eagerly caught at and retained by private Masters.[97]

Discontent over wages became chronic and threatening after the Seven Years War; overtime was discontinued altogether just as the cost of living, flat since the beginning of the century, started a long, although not particularly steep, ascent;[98] the intensity of labor slackened, discipline broke down, and the shipwrights, supported by the master shipwrights and the Navy Board, petitioned the Admiralty again and again for a pay rise. It was against this background that the Navy Board in 1765 cried up the benefits of overtime and the Admiralty in 1767 became alarmed about "Disorders and Irregularities in the yards". The Admiralty did restore overtime in 1765 but subsequently curtailed it ever more sharply year after year. In 1769 a common petition for overtime in summer, and a pay rise as well, was organized by the Plymouth shipwrights. The request for a rise and the common petition were unprecedented. The petition, having failed, was renewed in 1772 and failed again.[99] A pay rise would have required an order-in-council, however, and was out of the question; other groups of workers would not have been far behind with their own demands nor (an even more alarming prospect) might the Navy's wretchedly paid seamen; there was, indeed, no telling where it might end. The political nation expected cheap government, and the proper subordination of the lower orders, and was not disappointed. Task work, because it increased wages substantially, yet left the underlying time-rate undisturbed and at the same time eliminated overtime, was an irresistible solution to an intractable problem.

The innovation of task work in the dockyards in 1775 Sandwich considered to be his greatest achievement as First Lord.[100] The arguments in favor of task work were certainly powerful: more ships could be built with fewer shipwrights, dependence on commercial shipyards would be diminished, the least productive shipwrights could be discharged and the wages of the rest would be competitive with those paid in London. It was out of the question for financial and disciplinary reasons, however, that the prices should be a subject of negotiation; thus they were set unilaterally by the Admiralty. The Surveyor, Sir John Williams, with great difficulty drew up a scheme which divided ships into twenty-five stages of construction, called sections, "from laying their Keels to their launching," and established a price for each class of ship and for each of its sections.[101]

The shipwrights were understandably uneasy about the way task work would be administered; their concerns were largely laid to rest, however.

Men who were either old or lazy were excluded from the task gangs (there were 95), and the gangs were to work task in rotation (task work was for building and major repairs; time-rates continued to be paid for everything else).[102]

Task work commenced, on a trail basis in April 1775, on ten ships that were still in the early stages of construction.[103] The shipwrights appeared to be taking to it readily enough when Sandwich set out on his annual excursion to the dockyards in June. Early on the morning of 14 June, however, as he was preparing to disembark at Portsmouth, most of the task men in the yard (315 out of a total of 422 shipwrights) complained of "the hardship of being obliged to Work Task at the present price, and...left the Yard in a tumultuous Manner in a body, declaring that they would not return...'til Task work was abolished." The strike engulfed all the yards save Deptford and Sheerness and, sustained by high emotions, intimidation and violence, lasted until August. At its height, rather more than half of all the shipwrights were out, mostly task workers but a considerable number of sympathetic time-workers as well. Starvation and the indictment of the ringleaders at Woolwich for conspiracy to procure an increase of wages (a Common Law offence) finally broke the strike. Chatham's shipwrights on 21 August were the last to return to work. Not all the strikers were taken back, though; an example was made of 203 (including 45 who refused to return and 32 apprentices), and these were banished from the yards forever, although 74 old men were soon given a reprieve.[104]

Sandwich agreed that the prices were too low and immediately gave instructions to review and revise them "without delay." Most shipwrights working task had been earning 4s.-4s. 5d per day, i.e. 20-30 per cent more than effective time-rate of 3s. 4d. Sandwich was mistaken, however, in thinking that the promise of higher earnings would bring the strikers to accept task work. They rejected the premiss that higher pay should be conditional on higher productivity and wanted not just the abolition of task work but an increase in the time-rates as well.[105]

Naturally, the time-rates were not increased, and task work was not abolished, although it was made voluntary. Once the prices were revised, earnings rose on average to 5s. 3d. per day (26 per cent more than the highest time rate). There were few volunteers, however, except at the small eastern yards. Those who did volunteer, it is clear, were the most productive men.[106]

The War of American Independence followed and with it a severe shortage of shipwrights. The dissatisfaction of the day gangs continued to soar; indeed, it was at times only with the greatest difficulty that the day gangs could be persuaded to do any work at all. Finally, the master shipwrights were permitted to improvise piecework for repairs. This consisted of various "jobs"[107] each of which was some part of an article of task work; the prices were extrapolated from the task scheme, and the quartermen ascertained what work had been performed. It was expected that piecework would be used sparingly, and the prices had to be approved by the Navy Board; nevertheless, the master shipwrights were given a pretty free hand and paid the men as much as seemed to be necessary.[108]

Piecework (although not task work) was discontinued, and the time wage was reduced, either to 2s. 8½d (summer) or 2s. 1d. (winter), once the war was over. This was naturally a severe shock to the shipwrights, especially since the cost of living was then higher than at any time since before the war, and the London builders were paying more than double dockyard wages. The shipwrights petitioned for the restoration of piecework; indeed, there were even petitions for task work from men who only a few years before had sworn never to accept it. The master shipwrights advised the Navy Board that piecework must be restored. They had been guilty, however, of paying excessive prices, so a ceiling was put on earnings (either 3s. 4d. or 2s. 8½d. per day, depending on the season). Piecework with limited earnings, as it was called, was actually just a novel subterfuge (preferable to overtime) for paying a higher time rate and might better be described as day pay with check measurement, since wages in principle were to be checked when earnings fell below the limit. It was more and more restricted, though, as the need to gratify the men receded.[109]

The relationship between work and authority was delicate and complex. The men as a rule either ignored or defied those regulations which did not accord with their interests, and the officers as a rule acquiesced. A number of influences were conducive to this situation, among them low wages (and the perception by the yard officers that they were low), the shortage of artificers in wartime, and the fact that Britain was at war more often than not between 1689 and 1814. The social and cultural environment, however, was a subtler and more intriguing influence.

The shipwright communities were very tight-knit and had a well-developed ethos. This culture was reinforced by the developing "hereditary" character of the dockyard service and the emergence of extensive "clans" based on blood, marriage, business dealings, or political interest.[110] Most supervisors came from this same milieu and might themselves belong to one of the "clans." In any case, they lived amongst the workmen and were certain to be treated roughly if they thwarted the men's interests. Even the master shipwrights and other officers shared the shipwright culture to some extent. Had they not during those formative years when they were apprentices, and for several years thereafter, worked shoulder to shoulder with ordinary shipwrights?

The abuse of the privilege of chips, whereby the men were permitted to collect and take home small bundles of wood chips to be used as firewood, encapsulated the dockyard culture and defined the limits of authority.[111] There was a ready market locally for chips which the men were not at all backward in supplying. Thus valuable timber and plank were cut up just to make chips, and huge bundles, and even top ends and slabs, were carried out of the yard on the shoulder, in full view of the principal officers, who were required to station themselves at the gate precisely to prevent such things from happening. Middleton in 1783 issued an order that chips were not to be carried out of the yard on the shoulder. At Portsmouth, however, "the Artificers...in a body refused complying with it, and would only lower them from the shoulder to the Arm in passing the Gate, and their determination seems preconcerted." The officers turned a blind eye; to have done otherwise might have brought a complete breakdown of discipline. A supervisor — a man of "extraordinary...zeal for the public service, and a superior sense of duty"—who unwisely tried "to check in some degree" the making of illegal chips, had for his own safety to be escorted by a guard on his way to and from the dockyard. The intimidation of unpopular supervisors by means of graffiti, anonymous threats and physical assaults continued, indeed, well into the nineteenth century.[112]

The shipwrights felt that scrounging for chips, and making illegal chips, was demeaning and deeply resented having to do so in order to support their families. In 1783 they petitioned to have the privilege commuted to wages (an additional 5d. per week) and were supported by the yard officers. Indeed, the brazen defiance of Middleton's order soon after was undoubtedly intended to make a point. The Navy Board advised the Admiralty that the cost would be "trebly repaid", yet another eighteen years passed before chips were commuted.[113]

One might think that the materiel of naval construction, being inanimate, would have been easier to manage than people, but it was not. There were two categories of stores differentiated by use: dockyard stores and sea stores. It is with the former that we are concerned, although both were mostly the same. Nearly all stores were obtained by standing contracts negotiated by the Navy Board; in fact, standing contracts were the rule until the 1870s, when they were largely replaced by annual contracts.[114]   Our interest, however, is in the planning necessary for maintaining inventories at levels sufficient for production and, especially, in inventory control. The one was, unfortunately, highly tenuous precisely because the other was so muddled.

The information available for planning was considerable but mostly useless or, at best, misleading. The Surveyor of the Navy submitted quarterly to the Navy Board an estimate of how much additional stores of all kinds each yard needed for the ensuing quarter, whereupon the Board issued the requisite orders, and the stores were delivered to each yard direct by the contractors, from Deptford yard (the general depot for stores), or from a yard that was overstocked. The estimate was based on the yard storekeeper's quarterly returns of stores issued and remaining in stock. These returns, however, were often in arrears — as much as three-and-a-half years in 1749 and five years in 1802 — and were unreliable in any case. The storekeeper's monthly issues returns were sent to the Navy Office also, but these too were unreliable and, in fact, were only used by the clerks as scrap paper. Planning based on information of such quality was bound to be largely ineffective. Thus responsibility for keeping stocks up fell, in practice, upon the dockyards themselves: when shortages threatened, the principal officers collectively submitted a requisition, called an "occasional demand", to the Surveyor. Occasional demands were originally intended to be an extraordinary means of obtaining stores but became the ordinary means instead. The Navy Board was deluged with such requisitions in wartime, but the officers did not always act in time, and production might be held up as a result. That was sometimes the fault of the officers themselves, but so weak were coordination and inventory control that they were often caught unawares.[115]

The problem of inventory control began when the stores were received at the yard. It was the personal and (so as to prevent fraud) joint responsibility of the principal officers to survey the stores and certify that the quantity and quality were as specified in the contract, whereupon the stores were debited to the storekeeper. The officers were occupied, however, with much else that was more important and sensibly delegated the verification of quantity (and in some cases quality) to their clerks, for the procedures, although complicated, [116] were nevertheless routine.[117] The clerks, however, were not supervised and the officers accepted no responsibility for their actions. The contractors, in order to get prompt attention, supplemented the clerks' pitifully small salaries (starting at £30 per annum) with certain customary fees and gratuities the sum of which actually exceeded the clerks' official pay. Such payments were expressly forbidden but winked at anyway (they spared the government having to raise salaries). They were not bribes; nevertheless, they created a conflict of interest, and although most clerks were upright, many were only too willing to oblige dishonest contractors. Middleton was powerless to increase salaries and abolish fees and gratuities but did try, although unsuccessfully, to reform the system. The clerks were required to examine and countersign each other's accounts and the principal officers to certify them. These, however, were mere paper checks and of no practical effect.[118]

The official procedure for requisitioning stores was designed not for expedition but to prevent misappropriation. Stores sufficient for the work in hand were, sensibly, kept in current-use store cabins which it was the master shipwright's responsibility to keep supplied. However, his requisitions had to be countersigned by the clerk of the cheque, as well as by the storekeeper, and all three officers kept separate sets of accounts as a check on each other. Planning and coordination were not very good, and stores often ran out at the work sites. When that happened, the quarterman went to the foreman of the yard for a demand note which he then took to the master shipwright, the clerk of the survey, the storekeeper and, finally, the cabin-keeper. (In wartime, the master shipwright might spend several hours a day scrutinizing and signing such notes.) So circuitous was this procedure that it was frequently bypassed, the quarterman (or even a workman) going straight to the current-use cabin. There was, however, no good system for keeping the cabins stocked, and the stores might have to be fetched from a storehouse some distance away.[119]

Full and accurate accounts are the key to inventory control. Yet none of the criteria for control was met; this was basically why the dockyards were so often under- or oversupplied or supplied with the wrong species of materials. First of all, it is essential that materials which have been received should actually be brought to account. This was not the case, however, if clerks colluded with contractors and falsified quantities. Second, statements of quantities issued from inventory must be supported by properly authorized and receipted requisitions. No use was made of the receipts, however. Separate accounts of the quantities and value of all materials issued in the authorized manner were kept by the three departments involved in the requisitioning process; each month the master shipwright's and clerk of the survey's clerks compared their accounts and the storekeeper's monthly issues account and warranted the accuracy of the latter, whereupon it was delivered to the Surveyor of the Navy. There was no standard accounting procedure, however, and the accounts themselves were carelessly compiled and arbitrarily reconciled when they disagreed (as they frequently did). Moreover, the cabinkeepers kept no accounts at all (an invitation to misappropriation), and stores issued to them, and unexpended at the end of the month, simply disappeared without a trace from the storekeeper's monthly debit account. Thus it was impossible ever to know how much inventory was actually on hand. Finally, the balances of stores assumed to be remaining in stock must be verified by the actual taking of inventory. Inventory was taken, however, only upon a change of storekeepers (and not even then in the case of timber), which normally occurred only when the incumbent died (after an astonishing forty years in office in one case), and then only because the storekeeper was personally charged with the materials and had to pay a large bond. The argument against taking general inventory more frequently was that it was time consuming (it took three or four months), expensive and disruptive; this, however, was only because it was ill-planned and ill-conducted; otherwise it could have been done in three or four weeks. [120]

Middleton in 1778-79 introduced several remarkable, but unsuccessful, measures to facilitate planning, ensure an adequate supply of stores, tighten inventory control, establish cost control, and prevent embezzlement (workmen were no longer to draw cabin stores). The master shipwright estimated the week's stores requirements in advance, the foremen of the yard estimated the quantities the cabins would require, and the cabinkeepers requisitioned the stores on the Monday. The cabinkeepers were now charged with the stores and recorded both the issues and the works to which the stores were applied. The master shipwright, assistant master shipwrights and clerk of the survey at the end of the week audited the cabinkeepers' accounts by comparing them with the stores actually remaining in inventory and at the end of the month provided the Navy Board with an account of stores both received and issued and of the services for which they had been used. All of this, in the midst of a great war, was to pile insupportable additional work on officials who were overstretched already, and the new regulations were consequently ignored. A companion measure, designed to provide the Navy Board with a continuous overview of stores expended on each service, and as an aid in estimating long-term requirements, only complicated the system of accounts even more without achieving either purpose. It is sad to say, but Middleton's good intentions and ingenuity achieved nothing at all; it was to be a very long time, however, before anyone else succeeded in doing better; seventy years on there was still no accurate record of the quantities of stores that were actually consumed.[121]

Timber was of all stores the most important and also the most intractable and worst managed. Enormous quantities were wasted, and critical shortages were a not uncommon occurrence. Because no effort was made until the 1770s to acquire an adequate reserve of seasoned timber in peacetime, many ships had to be built of green timber in wartime. Such ships had to undergo a major repair after only six or seven years of service, compared to thirty years for ships built of seasoned timber, thereby placing an additional burden on the dockyards, diverting labor from new construction, and contributing to the Admiralty's dependence upon commercial builders. Not until 1753 did the Navy Board acknowledge the necessity of maintaining a minimum reserve (thought at the time to be equal to two-and-a-half years normal consumption), yet nothing further was done. One-and-a-half years supply was then either on hand or due on contract; nor was the situation any better in 1771; the western yards had at that time one-and-a-half years reserve at most and the eastern yards even less. It was in 1771 that the Admiralty adopted the policy of keeping a three-year reserve, calculated at the time to be 66,000 loads (it should actually have been 78,000).[122] This was acquired in just a few years and, despite shortages anyway, saw the Navy through the War of American Independence, although only just (Chatham in 1784 had a six-months supply).[123]

The timber reserve was undeniably a big step forward; in other respects, however, timber management remained as deficient as ever. The timber, because of inefficient systems of distribution and control, was unevenly distributed. The western yards, being near the forests, had an excess of timber, while the eastern yards had much less than their quotas. Moreover, a yard often received the wrong kinds of timber. Timber tendered the Navy Board was inspected by a purveyor (a shipwright specializing in that kind of work), but he was never told what sizes the receiving yards needed, and the timber was, as a rule, carelessly surveyed upon delivery. For example, it was often compared either superficially or not at all with the purveyor's report, thereby enabling the contractor to unload a disproportionate quantity of small

pieces for which there was relatively little use and which could not have
readily been disposed of anywhere else. The acceptance (albeit at a
reduced price) of small timber was eventually regularized (the contractors
insisted upon it as a condition of doing business) and continued down to
the 1860s, when iron largely replaced wood.[124]

The policy of maintaining a three-years reserve of timber created a
storage problem. Newly received timber was seasoned for one year in
open sheds, and more and more sheds were now erected. The problem
arose after the first year.   Some yards, simply overwhelmed with
seasoned timber, stacked it higgledy-piggledy wherever a bit of space
could be found, sometimes fifteen or twenty feet high and without proper
ventilation, so that much of it spoiled. Often when a particular piece was
wanted it could only be extracted after shifting many other pieces and
then might have to be dragged half the length of the yard; meanwhile, the
shipwrights who needed it might be standing idle. The economical
conversion of timber from a rough to a sided state, so that what was left
over was useable, was very exacting, so much so that even the specially
trained timber converters did not always succeed.   Nevertheless,
conversions were often performed by untrained quartermen and
shipwrights more intent on making chips than anything else. Moreover,
timber was stored untended and accessible to all and sundry, so there was
no telling how much was pilfered and cut up to make chips.[125]

The yards required large and continuous infusions of fixed capital if
they were to serve the needs of a navy that was growing by leaps and
bounds.   When Egmont visited the yards in 1764, however, he
everywhere found evidence of neglect and, in many cases, decay. "From
the want of sufficient and proper Docks and Slips, and fit and durable
buildings, the works cannot be carried on with advantage or preserved
as they ought to be." Neither Portsmouth nor Plymouth was equal to the
demands put upon it, while Chatham was in a ruinous state. Plans for
extending and improving the two western yards had been drawn up
already in 1760-61, but Egmont thought they were too modest and that
the eastern yards should be included. His instructions to the Navy Board
to draw up a long-term plan for each yard are remarkable as the first
example of comprehensive, long-range planning for the dockyards.[126]

Civil architecture and engineering were the province of the Surveyor
and the master shipwrights.   Thus it was Sir Thomas Slade, the
Surveyor, who drew up the plans, but only for Portsmouth, Plymouth
and Sheerness. Nothing was to be done at the other yards, because of

their declining importance and the staggering cost of the work actually proposed (estimated at £731,000 for the western yards alone). Portsmouth and Plymouth were crucially important, because of the westward shift of the foci of war, for small repairs and maintenance. Slade's plan for Sheerness, a small, make-shift but (because of its location) potentially very important yard, was not executed because wood-worm was such a problem there and copper-sheathing was still in the future. Work at the western yards commenced in 1765 but proceeded very slowly, although according to plan, until Sandwich attacked it with characteristic vigor; it was almost entirely finished by 1786.[127]

Portsmouth and Plymouth were transformed, becoming, as Sandwich observed, "superior to any thing in Europe." The changes were rather more striking at Plymouth; it was at last "a complete Dock Yard, such as it ought to be," rivalling "if it did not exceed Portsmouth in many particulars". Sandwich thought, however, that some of what Slade had planned for Plymouth was nothing but "useless Magnificence." These projects were abandoned, and the money thus saved was spent instead on improvements to Deptford, Woolwich and Sheerness and, more important, on providing nine additional building slips for frigates (two each at Portsmouth, Plymouth, Chatham and Deptford and one at Sheerness). It was because of the vast construction projects of the 1760s and 1770s, and later after 1795, that Middleton, for once content, was able to say in 1804 that the dockyards were in "pretty good order".[128]

Layout was another matter. Melville observed in 1813 that the dockyards had not been laid out with a view to their purpose but consisted of "a Succession of Make-Shifts", a stricture from which, however, he notably excepted Chatham. Poor layout was the inevitable consequence of sporadic, unplanned growth and the Surveyor's and the master shipwrights' lack of training in civil engineering (a profession which only originated in the later eighteenth century), so that even a dockyard wholly built to a master plan might still have been ill-arranged. Indeed, Slade's unexecuted plan for a new dockyard at Sheerness was forty years later pronounced deficient by John Rennie, one of the most eminent civil engineers of his day and author of the plan that was eventually adopted for that yard. But not even a country as wealthy as Great Britain could afford to scrap existing naval yards and build anew. Most of the dockyards in the age of iron and steel ships were basically the dockyards of the eighteenth century, condemning naval constructors in the late nineteenth century to cope as best they could.[129] Yet the dockyards in 1914 were probably laid out no worse than most big commercial yards, including some of the so-called "battleship" yards that built for the British and foreign navies.[130]

The Royal Dockyards in the eighteenth century were inefficient, and production costs were high; planning, organization, direction, coordination and control were feeble. It was essentially the system that was at fault. The Navy Board attempted the impossible task of micro-managing the yards as if they themselves were actually on the spot. Information resources were practically useless; there was no control over finances, costs, or inventory. The master shipwrights, unlike the men who managed commercial shipyards, were not businessmen but defined themselves instead as traditional craftsmen and took a craftsman's narrow view of their responsibilities. Nor had they received an education appropriate to the high managerial functions they were called upon to perform. They were given no latitude to take decisions and consequently held themselves responsible for no more than carrying out their literal instructions and were little concerned with controlling costs. The multifarious administrative duties associated with large, complex organizations were collectivized, on the one hand heaping upon the master shipwrights more duties than they could humanly perform, and in that way blunting individual responsibility, and on the other hand making it easy for all to shift responsibility and shield themselves from blame. Authority was undermined by a corrupt system of appointments , while dissatisfaction with wages made it almost impossible to increase labor-productivity and a constant struggle just to hold worker-output level.

The management of the dockyards was rather like the standing orders, the result of unplanned, incremental growth and for that reason deficient in method and order. It was the conspicuous want of order, noted again and again in the records of the visitations, which the reformers of the later eighteenth century attacked; they never questioned the underlying system of organization and management, however; Middleton, in fact, hardened it. Nor did the reforms amount to all that much in the long run; the dockyards at the end of the eighteenth century were still being managed, and were still burdened with the same problems, as they had been when it began. Just now, however, the opportunity for systemic reform presented itself.

Chapter 3

Tinkering with the System, 1793-1815

An extraordinary thing happened around the turn of the century: the dockyards became a subject of parliamentary and public interest and, indeed, political controversy. There were reports by the Commission on Fees (1788), the House of Commons Select Committee on Finance (1798), the Commission of Naval Enquiry (1803) and the Commission of Naval Revision (1805). Meanwhile, a management consultant was appointed (1796). The dockyard organization was shaken to its foundations. The process of systematic reform had begun; however, systemic reform (an altogether different thing) was made to wait quite awhile longer.

Parliament until now had shown no interest in the inner workings of the various departments of government and was content to leave administration (and the spending of money) to the ministers of the Crown and their subordinates. During the War of American Independence, however, the parliamentary opposition attacked corruption in government; Lord North responded with the Commission of Public Accounts (1780) which in its reports enunciated "principles fundamental to the subsequent improvement and modernization of public service."[1] One of its recommendations led William Pitt in 1785 to establish the Commission on the Fees, Gratuities, Perquisites and Emoluments in Public Offices with a view to abolishing all such. The Commission naturally delved into the organization and management of the Admiralty, Navy Office and dockyards, and, Middleton having gotten its ear, followed his recommendations in respect of the last two.[2] However, Pitt, fearful of how powerful vested interests, and his own parliamentary allies, might react, sat on the reports, and Middleton consequently resigned the Comptrollership (1790). It was because of the determination of Earl Spencer (First Lord from 1794 to 1801 and an admirer of Middleton)[3] that the report on the Navy Office was finally implemented in 1796. The Navy Board's business was now distributed among three committees and the commissioners were relieved of departmental responsibilities.[4] Meanwhile, Great Britain since 1793 was at war with revolutionary France, and the dockyards were stretched as never before.

Enter Samuel Bentham (Jeremy's younger brother), the management consultant and unquestionably the most remarkable person ever associated with the dockyards.[5] As a boy preparing for university at Westminster School, he persuaded his reluctant father (a lawyer) to let him make his career in the dockyards and was accordingly apprenticed in 1771 to the master shipwright at Woolwich. His ambition at this time was to be Surveyor of the Navy; in order to qualify himself pre-eminently, he was by prior agreement given time from the usual routine to obtain an education such as no other apprentice of the superior class had ever yet received, studying, among other subjects, geometry, chemistry, electricity and mechanics, and being instructed by a mathematics master; since most works on marine architecture were in French (none was in English), he went to study French at Caen in 1775. With Sandwich's approval, after his apprenticeship he attended the Royal Naval Academy, spent two months at sea, and studied naval construction at various British and foreign yards. Meanwhile, he despaired of ever becoming Surveyor. Thus in 1780 he went to Russia, where he designed, built and managed a small naval yard for his patron Prince Potemkin, was commissioned to make a survey of the country's mining and metallurgical industries, became a brigadier-general and was knighted by Catherine the Great.[6]

Bentham returned home from his Russian triumphs in 1791 and soon propelled himself into the most remarkable period of an unusual career. He undertook a survey of British manufacturing and continued experiments (begun in Russia) with steam-powered woodworking machinery, using these to bring himself to the attention of the Admiralty in 1794. The Navy Board was full of doubt about mechanization, but so intrigued was Spencer that he gave Bentham permission to visit the yards and report on the extent to which it might be adopted. Bentham went further, however, heaping devastating criticism on the ambitious improvements then being executed at Portsmouth and drawing up the alternative plan (he claimed that it would treble the yard's capacity) which now abruptly superseded the Navy Board's own plan. Next, he persuaded Spencer (whose confidence in the Navy Board was already shaken) to create for him the office of Inspector-General of the Navy (March 1796).[7] It was Bentham's job as Inspector-General to advise the Board of Admiralty (to whom he was directly and solely responsible, thereby bypassing the Navy Board) on civil construction, materiel and, by extension, the management of the dockyards. Thus were Bentham and the Navy Board at odds from the moment of his first appearance. Nor were their relations helped by Bentham's (like Middleton's) high self-regard and low opinion of the Board.[8]

Bentham is above all important as one of the very first management theorists. The dockyards were ill-managed, he argued, not because of the individuals concerned but because of a system of organization based on unsound principles. There were, he said, three sovereign principles of sound management, all three of which were negated in the dockyards. These were clear lines of authority, individual responsibility, and full and accurate information about costs, operations and materiel. Individual responsibility was the key, however.[9] It was, indeed, this novel idea, taken up and propagated by his famous elder brother Jeremy,[10] that was to dominate the thinking of nineteenth-century administrators. Samuel was not given the opportunity to apply his principles, except in one or two instances; the results might not have been successful even if he had been (he was inclined to overrate his abilities). Nonetheless, it was along the highly original lines he laid down that the management of the dockyards subsequently evolved, even if it did take a hundred years or so, and he himself in the meanwhile had been forgotten.[11]

Pitt in 1797 responded to the financial crisis brought on by the war with the monumental Select Committee on Finance. The committee's chairman, Charles Abbott,[12] happened to be Bentham's half-brother, and the Inspector-General's influence now rose even higher. It was, however, what the committee observed with their own eyes — an apparent breakdown of discipline and control in the dockyards far greater, they said, than what might have been expected in wartime — that prompted them to recommend the urgent necessity of a general visitation as soon as peace was restored. They were no less dismayed that the Commission on Fee's recommendations for the dockyards still had not been carried out.[13] Bentham was not wholly satisfied with these, however, and he and Sir Evan Nepean, the first secretary of the Admiralty, with Spencer's approval (but to Middleton's annoyance), now undertook to revise them.[14]

The recommendations of the Commission on Fees (as amended by Bentham) were issued by order-in-council in May 1801 and gave the dockyards a slightly more modern face. Apprentices were no longer a perquisite of office but were articled (in effect) to the Admiralty instead.[15] All "Fees, Perquisites, Premiums on the Appointment of Clerks and Emoluments of every kind whatsoever" were abolished, and salaries were increased in lieu thereof, thereby settling the problem of the clerks and the reception of stores. The foremen of the yard and quartermen were made salaried officials, thus assimilating them more to management, at least in principle. The privilege of chips was at last abolished in favor of an additional 6d. per week on wages,[16] and a new system of timber management, devised by Bentham, was introduced.[17]

Meanwhile, Bentham's views had involved him in a bitter dispute with the Comptroller, Sir A.S. Hammond (1794-1806), which became public in 1800 when Bentham published, with Spencer's permission, his *Answers to the Comptroller's Objections on the Subject of His Majesty's Dockyards*.

Spencer resigned, along with Pitt, over Ireland early in 1801. His successor, Admiral Earl St. Vincent, the hero of the great victory over the Spanish off the cape of the same name in 1797, was a literal-minded naval officer of inflexible opinions, towering Evangelical rectitude and merciless severity. The new First Lord, driven by the conviction that the entire administration of the Navy was a cesspool of corruption, soon declared war on the Navy Board.

Serious charges were brought against the foremen of caulkers at Plymouth, whereupon St. Vincent sent a committee down to investigate. The committee's report (February 1802) revealed, in the First Lord's view, such a degree of looseness and irregularity in the administration of piecework and overtime as to have defrauded the public of an incalculable sum of money.[18] St. Vincent sacked the master shipwright, storekeeper and clerk of the survey, as well as the foreman of caulkers, rejected the Navy Board's nominees to succeed the master shipwright as being "altogether incompetent to root out the abuses (to say no more of them)", and appointed his own man instead.[19] He was now convinced that nothing less than a commission of inquiry with a sweeping mandate would do and with this in mind impounded the records of all the principal officers and had them delivered to the Admiralty, where they were kept for nearly two years. The Peace of Amiens followed, and St. Vincent, faithful to the Select Committee's admonition, visited the dockyards, finding each one to be "a viler sink of corruption" than the last; Portsmouth was "bad enough," but Chatham beggared "all description." He accused the Navy Board of gross dereliction of duty and eventually broke off all private communication with Hammond.[20] Nevertheless, the minutes of the visitation suggest that conditions on the whole were no worse than after previous wars; there is, however, no point in recounting the innumerable irregularities and abuses that offended St. Vincent.

St. Vincent's reasons for wanting a commission of inquiry were almost wholly negative — punitive rather than reformative. Henry Addington, the Prime Minister, calculated the political risks and was opposed, but St. Vincent threatened to resign, and the Commission of Navy Enquiry was created by Act of Parliament in December 1802. Of the Commission's fourteen reports (1803-06), however, only two dealt with the dockyards.

It was perhaps inevitable that the Commission of Enquiry's sensational but misleading allegations of wholesale fraud and corruption should have political repercussions, dividing Parliament between those who enlisted in St. Vincent's vendetta and those who defended the Navy Board. Meanwhile, the First Lord flouted the Navy Board's advice and precipitately ran the fleet and the dockyards down after the Peace of Amiens; however, the peace unravelled in 1803, the ministry came under fierce attack, and Addington resigned (May 1804). Pitt was Prime Minister again with his friend the first Viscount Melville as First Lord of the Admiralty.[21]

Pitt, encouraged by Middleton, made up his mind to quell the tempest, so ravaging in its effects, that St. Vincent had created and to complete the job of reform begun by the Commission on Fees. In December 1804, therefore, he created the Commission for Revising and Digesting the Civil Affairs of the Navy. Middleton (now nearly eighty) was appropriately made chairman. Just now, however, the Commission of Enquiry accused Melville of malversation whilst Treasurer of the Navy years before, and his impeachment and resignation followed.[22] A peerage (as Lord Barham) persuaded Middleton to take on the additional burden of the Admiralty,[23] although he served as First Lord for but a few months (April 1805-February 1806), resigning upon Pitt's death.

Bentham, whose influence had waned, was sent to Russia (1805-07), not with the intention of getting rid of him but on an important mission superintending the building of much-needed frigates for the Royal Navy.[24] The Navy Board took advantage of his absence, however, and persuaded the Commission of Revision that the Inspector-General's office should be abolished (Middleton was probably pleased to oblige).

Bentham was not sacked but, ironically, was made a Commissioner of the Navy and given the title of Civil Engineer and Architect and considerably reduced responsibilities.[25] It was all downhill for him after that; his new office was abolished, and he was seen off with a handsome pension in 1812.[26]

Meanwhile, the Commission of Revision had completed its monumental work and dealt with the charges against the dockyards. These were, broadly, that the business was conducted in a disorderly manner, the accounts were inaccurate and the correspondence imprecise, and the officers did not understand their instructions. The Commission's accomplishments may be summarized briefly. Every yard official, from the resident commissioner on down, was provided with full instructions; the standing orders were revised and codified; time wages were increased; overtime was severely restricted; piecework was regularized; a yard exclusively for building was established; and the School of Naval Architecture was created. Thus had St. Vincent's witchhunt led to good after all. Bentham's principles of management were not embraced, however; the Commission of Revision, indeed, fully endorsed the principles laid down by Pepys in 1662 as having "stood the test of time" and so perpetuated and consolidated the existing system of organization and management. Individual responsibility, Bentham complained, was "*still a desideratum.*"[27]

The Commission's recommendations having been adopted (only a few were not), the Admiralty wanted to know if, and how faithfully, they were being carried out and how well they worked in practice. Thus it was that the yards were visited by the Navy Board in 1810 and the second Lord Melville in 1813-14.[28] As a result, certain further changes were made. A third surveyorship was created in 1813, and Robert Seppings, the master shipwright at Chatham (and one of the great naval architects of his time), was appointed. Part of Seppings's job was to visit the yards and make certain that the various innovations that had been introduced were properly understood and uniformly administered. The third surveyorship was a bold yet necessary measure; the office was unfortunately allowed to lapse when Seppings became junior Surveyor in 1822.[29]

The Commission of Revision was called upon to sort out the mess at the Navy Office resulting from the reorganization of 1796. The objects of that measure (inspired by Middleton) were greater expedition and order in conducting the Navy's Board's business and institutionalization of the Comptroller's de facto presidency. The business was divided among three specialized committees, and "a general superintendency and directing Power" was created by making the Comptroller chairman of both the Board and its committees and giving him financial oversight. The Committee of Correspondence (Deputy Comptroller,[30] Senior Surveyor and one junior commissioner) was responsible for managing the dockyards and for all other correspondence as well. The other committees were the committee of (financial) Accounts (three junior commissioners) and the Committee of Stores, including storekeepers' accounts (Junior Surveyor and two junior commissioners).[31] The actions of the committees were subject, however, to ratification by the full Board, and the Board alone had jurisdiction over contracts. Responsibility for the staff and for the administrative routine of the various departments was shifted from the commissioners (so as to free them to function strictly as managers) to their chief clerks; at the same time, several new departments were created, most notably a separate department of stores, a measure which was to create many problems when the Navy Board was abolished.[32]

The new structure of the Navy Board was immediately put to the test of war and failed. Business grew by leaps and bound, that of the Surveyor's and the store departments doubling between 1796 and 1806; everything fell more and more into arrears, arrears of accounts being greater in 1806 than after the War of American Independence; nor does the appointment of secretaries to the Board and the committees seem to have produced much greater order in the way business was conducted. The commissioners were collectively responsible for the departments under their committees and thus inclined to shift final responsibility to their colleagues; moreover, difficult questions had to be referred to the full Board. The Comptroller was expected to be the linch-pin, but he, even more overstretched than his colleagues, seldom presided over the committees. Business was still conducted under pressure, in haste, and often in a muddle; to the clerks, underpaid and overworked (pleas for more clerks were rebuffed), was left the task of giving final shape to the Board's decisions. No wonder that "errors, irregularities, and inaccuracies" often crept into the orders that were given the yard officers. The compromise solution, adopted on the recommendation of the Commission of Revision, was to make each commissioner of the navy responsible once again for a particular part of the business[33] and to relieve the Comptroller of being chairman of the committees.[34] It was still a bad system, however; yet it lasted until 1829.

The way the dockyards were managed hardly changed at all. Bentham said that the yard commissioner should exercise "absolute power" over "the whole *operative* business..., making him manifestly responsible for the [general] *business*" of each department but entrusting the details of management to the principal officer. All (the commissioner and the principal officers) should be held "individually responsible" for their conduct; no longer should the execution of orders be entrusted "to a number of persons in the aggregate, on no one of whom can individual responsibility be affixed." Indeed, the order in council of 1801 defining the commissioner's duties asserted that it was "essential, that in every Case, some One of the Officers, should be made to stand individually responsible for the due execution of the [Navy Board's] Orders". The commissioner was now explicitly given "full Authority" but only for the purpose of "enforcing Obedience" to the Navy Board's orders and regulations and of "obliging every person to discharge the Duties of his Office...[and] correcting Abuses," and so on, all of which the Commission of Revision repeated when it revised the dockyard regulations. The Commission, however, made certain additions which, it thought, gave the yard commissioner greater weight. Henceforth, the Navy Board's letters were sent to him first (they continued, nevertheless, to be addressed to the principal officers), as were the officers' letters to the Board (the commissioner might write comments either in the margin or in a covering letter). Finally, he himself wrote to the Board daily, repeating "the heads of every letter, and order addressed by the Board to the Officers," so as to satisfy the Board that he had "a knowledge of every subject." The commissioner was expected to spend most of the day prowling the yard and making sure that the Board's orders were actually being observed and that everyone, from the officers on down, was doing his duty. He was empowered to suspend and recommend the dismissal of malefactors, but the power to dismiss was explicitly reserved to the Navy Board (to the Admiralty in the case of officers). These were the functions not of a manager but of a postman and policeman combined in one person. The commissioners, moreover, continued to be appointed as a reward for past services and without much regard to aptitude.[35]

Work discipline was the biggest challenge facing the dockyards during the wars, for the shortage of shipwrights and certain other artificers was persistent and at times acute,[36] the London builders in 1794 were forced to raise the time-rate to 5s. 3d. per day,[37] and the price of foodstuffs simply ran away in the late 1790s before levelling off after 1801.[38] An increase in the time rates was still out of the question, however, and the Navy Board reached into its bag of tricks. Limited piecework, with a limit of 4s. 2s. the year round, was quickly revived, only now it was incongruously, yet quite frequently, supplemented with various amounts of overtime, the shipwrights sometimes earning as much as 6s. 3d. per day. Soon all the artificers were being paid in this same way. The prices were purposely not divulged, however, and as a result the value of the work generally exceeded the limit. When the shipwrights at one yard were informed of the prices, they refused to perform more work than they could be paid for.[39]

There were at first no standard piecework rates; each master shipwright was more or less free to decide for himself how much to pay. The master shipwright proposed to the Surveyor a price for each species of work when a ship was to be repaired, and the prices actually authorized by the Navy Board were to apply at his yard thereafter; meanwhile, prices varied (often markedly) from yard to yard. The procedure adopted was sensible enough in principle, for piece-rates can as a rule be developed only on the basis of data accumulated over a long period and even then must frequently be reviewed; especially is this true of non-repetitive work like shipbuilding; moreover, repairs, involving assembly operations to a much lesser degree than building, present particularly difficult problems of pricing. The limit on earnings and the allowance of overtime were impediments, however, to ascertaining how much work the men were actually capable of performing in a standard working day. Nevertheless, a number of standard rates evolved (through the process of review in the Surveyor's office), and in 1802 a schedule consisting of the principal articles (about 10 per cent of the whole) was drawn up. The rates were defective, however, and earnings varied widely, between 3s. 4d. and 7s. 8d. per day.[40]

Piecework with limited earnings plus overtime allayed discontent and kept the men at work (although not necessarily hard at work), but only for awhile. For discontent, coinciding with the final, spectacular surge of price inflation, food riots and fresh (although unsuccessful) wage demands in London,[41] broke out in 1801 with a ferocity, and on a scale, not known even in 1775. The whole body of dockyardmen petitioned for a pay rise and sent delegates to plead their case before the Board of Admiralty. The delegates returned to the yards apparently satisfied with a promise that bonuses would be paid while the cost of living was so high. The yards were struck anyway; indeed, they were laid under siege by men who would stop at almost nothing, and troops had to be summoned. Order was quickly restored, however, and the men returned to work. The ringleaders (311 men, including 128 shipwrights) were dismissed, but so critical was the shortage of artificers after 1803 that many were subsequently re-employed.[42]

The resumption of hostilities with France in 1803 necessitated furious preparations and a large increase in the workforce owing to St. Vincent's misguided policies. Thus the cap was taken off piece wages and overtime was discontinued. The yards continued to be in an agitated state, nonetheless, and it was a constant struggle to keep productivity from plunging. The piece rates were found to be excessive and were lowered by 19 per cent in 1804; however, the men five months later forced the cut to be rescinded. Task rates (now almost thirty years old) were so low relative to piece rates that it was only by paying the latter (and thereby increasing earnings by 45 per cent) that the task gangs at Sheerness could be persuaded to do any work at all. So many task men quit Chatham yard that the task rates applying to the ships building there had to be raised.[43] This was followed by a general increase of 20-25 per cent.[44]

St. Vincent, literal-minded and uncomprehending, thought that limited piecework and overtime, both singly and in tandem, were a huge swindle. Thus the question of wages could be evaded no longer. It was clear to both the Commission of Enquiry and the Commission of Revision that the time rates had to be increased. It was clear also that they should be higher in wartime than in peacetime. Thus in 1812 dual rates were adopted (e.g. either 4s. 6d. or 5s. for shipwrights in summer). Both Commissions discountenanced overtime, however, and its use subsequently was severely restricted in practice. The workers probably were barely aware of these changes, however, since very few were then being paid a time-rate.[45]

The two Commissions disagreed over task work and piecework. The Commission of Enquiry recommended scrapping piecework and paying a time-rate instead for all but large repairs (to be performed by the task). It was the Commission of Revision's argument that prevailed, however, leading to the institutionalization of piecework, the abandonment of task work, and more than fifty years of controversy at the end of which the Commission of Enquiry was vindicated.

Piecework was radically different from task work in that the hull was divided into all of its hundreds of constituent parts, thus making it necessary to measure the work as it advanced. The tables adopted in

1809, and consisting of about 2000 articles comprehending new work as well as repairs, were described thus by an awestruck Lord Melville.

> In new work, these tables commence with the first laying of the Keel of a Ship, and proceed from one piece of timber to another, from plank to plank,'till the ship is completely ready for launching; and in the old work every repair that can be wanted, is valued from the fitting in a beam, frame timber or knee, to the driving of a Bolt or treenail.[46]

The Commission of Enquiry doubted, though, that the articles could be measured accurately; many were "so trifling, as to render it difficult to take account of them, and to estimate the Value of the Labour performed." Moreover, the shipwrights were often suddenly shifted from one job to another, thereby rendering "the keeping of a faithful Account of the Earnings a very complicated Business", while the accounts themselves were, because of this and the character of the work, liable to errors and falsifications which were unlikely to be detected. There was also the high cost (quite apart from the time taken by the officers to supervise their compilation), of "the vast Number of Accounts required to protect the Public from Fraud...and...unintentional errors".[47]

The Commission of Revision approached the question of piecework with an open mind and let itself be guided by the experience of industry and the opinions of the master shipwrights. The Commissioners visited a number of shipyards and factories (including the works of Boulton and Watt) and found that:

> the practice of employing workmen by the piece, in preference to paying by the Day, [was] universal among them; it being the decided opinion among them that it is the most advantageous plan for both the Workmen themselves and those by whom they are employed.

The conclusion drawn from this was not just that a time wage encouraged idleness, and a piece wage the opposite, but that piecework automatically produced a high intensity of labor. The master shipwrights were consulted and advised the Commission that all, or nearly all, of shipwrights' work could be measured precisely. On the Commission's recommendation, therefore, a committee of master shipwrights was appointed and drew up new and comprehensive piecework tables for shipwrights' and certain other work.[48]

The Commission of Revision did not share the Commission of Enquiry's enthusiasm for task work. The thirty years since it was introduced had revealed a number of flaws partly reflecting the difficulty of pricing large stages of work satisfactorily. Small ships were priced too low and big ships too high (average earnings per day were between 3s. 1d. and 6s., depending on the size of the ship), while many of the sections of all classes of ships were mispriced also (e.g. the men earned on a 100-gun ship between 4s. and 7s. per day depending on the section). Since the prices in the task scheme only applied to specimen ships of specific length, breadth and depth[49], it was necessary to recalculate them (an exceedingly complicated mathematical exercise) whenever the dimensions were altered. This, however, increased the price disproportionately to the additional labor actually arising from the changes (e.g. another three feet on the width of a ship 196 feet in length added £900 to the price, even though the real value of the labor was only £30). It was decided that length alone was the best indicator of the amount of labor and to revise the prices accordingly.[50] A further problem was how to apportion the price when successive gangs worked on the same section. The Commission of Revision dealt with this by instructing the committee of master shipwrights to draw up a combined task/piecework scheme. Such, however, was the general disillusionment with task work that the new scheme (1809) provided for piecework alone, with the articles arranged by sections. Task work was now dead, except for the term itself, which continued almost until the end of the nineteenth century to be applied (misleadingly) to new work.[51]

The administration of piecework, originally merely improvised, very loose and wide open to fraud, improved enormously as a result of the recommendations of the Commission of Enquiry and, especially, the Commission of Revision. The quartermen up to this time measured their gangs' work and shared the gangs' earnings; their records were seldom inspected, however, and the work was valued, the accounts compiled, and so on, by shipwrights (seconded to the foreman of the yard's office) who shared their gangs' earnings. The quartermen in 1804 were put on salary, and in 1809 a master measurer, assisted by a professional clerical staff, was appointed at each yard. Contrary to the Commission of Revision's recommendation, however, only nine sub-measurers were appointed, to re-measure questionable work. The quartermen were now furnished with small blank books in which to copy printed descriptions of the work from a handbook; however, many quartermen were found to be incapable "of measuring work and rendering accounts thereof with sufficient accuracy for ascertaining the men's wages". In 1811, therefore, the quartermen were relieved of all measuring duties and the number of sub-measurers was increased to sixty-five. At the same time, the work of the master measurer's clerks was simplified, and its accuracy enhanced, by means of conversion tables for ascertaining the value of various quantities of any article. The man-hour content of the articles predictably remained quite elusive, however, and the expectation that the shipwrights would earn per day 6s. 3d. in summer and 5s. in winter was not realized. The prices were revised already in 1812 and afterwards were under continuous review, the yards from 1814 being required to report monthly all articles on which earnings either exceeded or fell below a certain maximum or minimum. Prices as a result were now changing constantly. So great was the work which this involved that it occupied as assistant surveyor of the navy full time. The shipwrights were content, however, and so they should have been, for they were earning on average 6s. 6½d. per day.[52]

The Commission of Revision thought that the piecework scheme would make it possible to compare the estimated and actual cost of repairs (the cost of new construction was still not estimated) and thus become an instrument of control. The scheme was not so used, however, and the attempt to do so would certainly have failed anyway, for as J.R. Parkinson has pointed out, it is not easy to integrate "payment-by-results schemes into an effective scheme of costing and budgetary control". For example, the "very mass of detail accumulated often appears to defeat its use in estimating" and as a result is not generally so used.[53] Nor were the dockyard piecework accounts sufficiently reliable to be of any use in controlling costs, as, indeed, Bentham observed already in 1798 but was not actually demonstrated until 1861.[54]

The Commission of Revision thought the shipwrights would turn out more work if they were permitted to choose their own workmates (as was done in private yards and in civil engineering, and, indeed, as the task gangs did already) and proposed introducing this method of selection experimentally at Plymouth.[55] The Surveyor and most of the master shipwrights were opposed, however, and argued (among other things) that it would give rise to "endless animosity, jealousy, and faction [amongst the men], whereas at present they are satisfied"; but it was the men themselves who, unfortunately, prevented the experiment from being made. Unfortunately, because it was to have been a controlled experiment comparing output per gang at Plymouth and some other yard and thus would have provided the historian with invaluable data about labor-productivity. Nevertheless, once the new method of recording earnings was adopted, and each gang was paid only for as much work as it actually performed, the younger and more energetic shipwrights (although perhaps not in every yard) insisted on rejecting old and worn out men, knowing, of course, that they would be provided for. The right of rejection lost its rationale, however, when limited piecework was revived after the war. The practice of reserving the best shipwrights for building was prohibited in 1833 but persisted at Plymouth until 1847.[56]

It was in this period that the weekly payment of wages replaced quarterly payments which, being so infrequent, obliged the men to go into debt and were a cause of much distress and discontent. The Navy Board made a connection between the delayed payment of wages and difficulties in recruiting shipwrights. Thus between 1812 and 1815 an ingenious and highly complicated method of calculating wages was introduced which made it possible to pay on account roughly what the shipwrights had earned the preceding week and to settle with them monthly.[57]

The Commission on Fees proposed that none but artificers (although only those who were "deserving, and diligent") should have apprentices. This, it was reasoned, would help to solve problems of work-discipline and subordination, recruitment and retention, and benefit the yards in another way also, since an artificer was likely to take greater care training a lad apprenticed to himself than one apprenticed to an officer. Bentham, however, persuaded the Admiralty that not all "deserving, and diligent" shipwrights were apt teachers of their craft and that apprentices should instead be articled to the master shipwright (representing the Admiralty) and assigned to qualified shipwright instructors. It was not an arrangement that worked out very well, though; the instructors were indifferently supervised, and many, even a century later, did not take their responsibilities very seriously.[58]

Another part of the apprentice system proposed by Bentham unfortunately was not implemented. Bentham thought the dockyards were poorly managed because the professional officers were ill-educated and would be managed even worse if the Commission on Fee's plan were adopted; no longer would an apprenticeship be eligible for well-educated boys of good family (those who were then apprenticed to the professional officers), and all the apprentices henceforth would be working-class and uneducated. Bentham wanted to establish at each yard a school where all apprentices would be educated according to their ability, with the brightest receiving a mathematical and scientific education (such as he himself had received) appropriate to a master shipwright. Pitt and Spencer gave their approval in 1800, but St. Vincent was more concerned with retrenchment, and the schools were not established when the new system of apprenticeship went into effect in 1801. The consequences were disastrous. The source of the old superior class of apprentices dried up at once; even many working-class parents no longer apprenticed their sons, although this appears to have been because apprentices were no longer permitted to share task or piece earnings. Apprentices after 1801 were mostly recruited from "the lowest class of people, and those least likely to have any education," and none was capable of receiving advanced practical instruction. So ignorant, indeed, were the new lot at Chatham that the shipwright officers took it upon themselves to designate a shipwright to instruct them (originally in nothing more than the three Rs), an arrangement which appears, however, to have been short-lived.[59]

The Commission of Revision impartially condemned both the old and the new system of apprenticeship; the one had not, and the other could not, turn out naval constructors fit to manage a dockyard. The one had provided an altogether inadequate education, while the other provided no education at all.[60] The Commission was no doubt aware of the proliferating academies that provided a scientific education useful to the rising class of manufacturers;[61] indeed, some dockyard constructors were themselves sensible, as one said, of "the disadvantages arising from the want of education as we advance in our profession."[62]

The Commission recommended creating a two-tier system of education — part-time apprentice schools, modelled on the one at Chatham, for the inferior class of apprentices and a School of Naval Architecture for the superior class, who were to be admitted by competitive examination, either after two or three years at an apprentice school or directly from public or grammar school. The apprentice schools were not established, but the School of Naval Architecture received its first students in 1811. The students in the morning pursued a rigorous mathematical and scientific curriculum[63] and in the afternoon received practical instruction, but, significantly, quite apart from the ordinary apprentices. Those who successfully completed the course were immediately appointed quartermen. Thirty candidates sat for twelve places in 1811, evidence of the eligibility among the status-conscious middle class of a professional career in the dockyards so long as the social context was right.[64] The School's graduates were highly unpopular, however, and the

School itself was shuttered after twenty-one years, long before its impact could be felt. Not until 1864 was it accepted that highly educated men were needed to design warships and manage the dockyards. Not until 1883 was the field of naval construction fully professionalized.

Bentham revolutionized the management of timber. In 1801 a timber master (a former assistant master shipwright) individually responsible for the timber was appointed at each yard, and an elaborate system of administration was created which was expected to provide order, method and economy where before there had been none. Of the appointment of the timber masters and the abolition of chips (both in 1801) the Navy Board committee that visited the yards in 1810 said, "Perhaps no regulations more beneficial in their effect were ever introduced into a Dockyard." Certain of Bentham's objectives were too ambitious, however, and the system as introduced was flawed by administrative anomalies, complexities and impracticalities. These were dealt with after 1810; nonetheless, fault was increasingly found with the system even as it was modified. Nothing was done, however, to improve the storage of timber, which continued to be as disorderly as ever and thus nullified much of the benefit anticipated from Bentham's system.[65]

Both Bentham and the Commission on Fees deplored the absence of inventory control; there was no uniform accounting procedure; the storekeepers did not keep accurate and up-to-date accounts and so were unable to balance their books quarterly. Portsmouth's practice of balancing debits (receipts) and credits (issues) quarterly and carrying the balances forward was adopted for the other yards in 1798 on the Commission's recommendation; however, the store accounts were in arrears until after 1813 (allegedly because of having been impounded by St. Vincent in 1802-04). The Commission's recommendation that stores should be surveyed annually and the results compared with the storekeepers' ledgers was adopted in 1801. "This method, uniformly pursued," the Commission thought, "will at least afford tolerable satisfactory evidence of the remains of all the principal articles, which at present is taken...upon trust". A general survey was taken the following year, but the war and the expense ruled out taking any more surveys. The Commission's recommendation that cabin stores (for current use) should be surveyed monthly was carried out also; however, the committee of visitation found in 1810 that these surveys were an insufficient check, since uniform procedures were not followed and the cabin-keepers juggled the books. Thus, additional checks, including double-entry bookkeeping, were introduced. The need (time-wasting and open to abuse) for quartermen to send to the cabins whenever their men ran out of materials was obviated, and at the same time better control was established over cabin issues, by providing the quartermen with chests for the safekeeping of materials that had been issued but were not wanted immediately. Despite these measures, however, there was still no very effective check on the expenditure of stores.[66]

The Navy in 1793 had nearly one hundred recently built or repaired ships of the line and so was unusually well prepared for war. These ships could not last indefinitely, however; they eventually needed to be repaired, and additional ships should either have been built as replacements or brought forward from the reserve. Yet remarkably few ships (certainly not enough) were built or repaired between 1793 and 1802. Perhaps it was not thought that the country would be at war as long as in fact it was to be; perhaps it was a matter of economy. In any case, this failure of policy eventually caught up with the Admiralty. The policy after 1805 (subsequently criticized by the second Lord Melville when he was First Lord) was to build as many ships as possible in preference to bringing them forward from the reserve. The dockyards as a result built twice as much tonnage between 1805 and 1813 as in the earlier period. This accounted, however, for only 13 (24 per cent) of the 55 ships of the line, and 23 (29 per cent) of the 80 frigates, that were built. Indeed, building accounted for very few shipwrights at any one time; it had to be sacrificed to refitting and consequently was interrupted again and again. That was not a very efficient way to build ships. Neither was it very economical; e.g. some work might be paid for twice, once when it was performed and again when construction resumed after an interruption.[67] The obvious solution, a yard dedicated to building alone, seems not to have occurred to anyone until quite late, however. A site was now chosen at Pembroke, and construction commenced in 1814.[68]

Refitting was a vastly more urgent problem. Streamline the yards and the problem will go away, Bentham said. This was the essence of his remarkable plan for Portsmouth which, when executed, even his rival and critic John Rennie acknowledged made that yard superior to any other.[69] What above all needed to be done at Portsmouth, Bentham told the Admiralty, was to shorten the turn-around time for graving and refitting. Thus, among his many projects the fitting basin was enlarged (to 274 acres); the entrance was deepened and, off the basin, two additional drydocks and three long jetties were built (to take advantage of deep water).   Bentham made large claims (not all of them well founded) for these improvements.   Before the basin was enlarged, ships often had to be refitted in the harbor; shipwrights and materials had to be ferried out, bad weather occasioned delays, and the shipwrights, since they could not be supervised away from the yard, were guilty of loafing, wasting materials and pilfering.   Now, however, the basin could accommodate one-third more ships (seven of the biggest), and refitting could be accomplished as it were within the yard itself.   Lowering the depth of the basin gates and extending the jetties into deep water made it possible to move ships in without having been lightened first (by unloading the guns, stores and provisions).   The new drydocks being deeper than the old ones, ships could be floated in and out at any time, without waiting for spring tides, thereby shortening the turn-around time in examining and cleaning their bottoms and making small repairs.   The biggest saving of time, Bentham thought, would be in refitting; only two or three days, instead of six weeks, would be required.   He was wrong; the turn-around time appears to have remained about the same, although not because it could not have been reduced (two or three days was a wild exaggeration, however), but partly because of poor management; work in the yard itself was constantly, and for the most part unnecessarily, being interrupted to send shipwrights to make small repairs to ships at Spithead.   The docking and undocking of ships was facilitated by using a steam pump to regulate the water level in the basin and docks, and a waterworks system was installed.   Bentham also developed the first bucket-ladder steam dredger, thus making it possible to clear the shoals at Portsmouth in the merest fraction of the time required to do so manually.[70]

Extensive improvements were underway at Plymouth also in the 1790s, but Bentham was much less critical of these.   For the eastern yards, however, the Navy Board still had no plans.   Nor did Bentham.   The value of Deptford, Woolwich and Chatham was too much diminished, he believed, by their distance from the sea, silting, and delays encountered in navigating the Thames and the Medway.   The useful life of these yards was prolonged nonetheless by his dredger.   Sheerness, even though highly useful for refitting, was in a ruinous state and badly situated. Bentham therefore recommended running the eastern yards down and produced a plan for a new yard on the Isle of Grain opposite Sheerness; however, the sheer cost sank the idea, just as a few years later it sank the Commission of Revision's recommendation to build an even more expensive yard, designed by Rennie, at Northfleet. The cheaper alternative of modernizing Chatham and rebuilding Sheerness (in both cases according to Rennie's plans, Bentham's plan for Sheerness having lost out) was adopted instead. It was from this time the Admiralty's policy to contract civil engineering out and place the works under the direction of a professional civil engineer.[71]

It was Bentham who brought the Industrial Revolution to the dockyards, despite the Navy Board's wariness. The steam-powered wood and metal mills that came into operation at Portsmouth in 1803-05 (unquestionably his greatest achievement) made the dockyards self-sufficient in a wide range of manufactured articles — wood blocks, copper sheathing, bolt staves and various kinds of castings. The machinery itself, however, was either invented or designed by Marc Isambard Brunel, who also supervised its erection and operation while being broken in, and it was Bentham's machinist assistant, Simon Goodrich, who actually designed the mills. Delays in bringing the mills on line according to schedule naturally created temporary supply problems, especially as the contractors who were being supplanted were not very cooperative. That was a small price to pay, however, for the enormous benefits the mills conferred. The wood mill by 1805 was manufacturing 150,000 blocks a year; now ten men could turn out as many blocks as 110 had done before; moreover, the blocks were superior to those the contractors had supplied and were manufactured at a saving of £18,000 a year, while copper sheathing was being remanufactured in 1812 at a saving of over £24,000 a year. No less important, all the engines and machinery could be fabricated in the yard's own shops. It must have been with a sense of great satisfaction and pride that Brunel showed off these impressive monuments to the new machine age when Melville visited Portsmouth in 1814.[72]

Bentham, being a believer in administrative unity, wanted to put the mills under the master shipwright, but the master shipwright (backed by the Navy Board) refused to concern himself in any way with their operation. They were transferred to his department anyway in 1849. Meanwhile, Goodrich was appointed engineer and machinist and given sole responsibility upon Bentham's departure from the Navy Board in 1812.[73]

The dockyards were an object of extraordinary interest and activity, and an unprecedented number of changes occurred, during the wars of the French Revolution and Napoleon. The Navy Board was reorganized, fees were abolished, a new system of timber management was introduced, the instructions and standing orders were revised and the latter codified, salaries and wages were increased, and piecework was systematized. The harsh glare of official scrutiny and criticism undoubtedly had a tonic effect on the Navy Board and put the yard officers more on their mettle also. It is Roger Morriss's "general impression" that although "the changes were individually of little consequence," collectively they constituted "a silent revolution"; because the wars "demanded an efficient dockyard service," by the time they came to an end "an already long-established and well-tried organization had become a smoothly humming instrument of war."[74] A "smoothly humming" organization, however, should not be mistaken as being a corollary of the need, however great, to be efficient; that is a conclusion that is impressionistic rather than evidentiary. Paul Webb, another student of the dockyards in this period, falls into the same error when he writes: "The most striking impression..., based as much on a 'feel' for the material as upon demonstrable proof, is that the Navy was remarkably well served by its building and repair policies." Yet the accounts which might have provided "demonstrable proof" are quite unreliable, as Webb himself admits. [75]

The fundamentals of sound management were no more present in 1815 than before. The system of organization was just as highly centralized and subversive of local initiative; the master shipwright was still responsible for no more than carrying out his literal instructions. A corps of highly educated professional officers prepared for their careers at the School of Naval Architecture was years away. There was still no general dockyard manager. Most sources of information were still unreliable. There was no financial control, cost control, or inventory control (except in the case of timber).

The system of organization and management was like an old banger of a car that the owner has fixed up; the car may wheeze less, shine more and make do for awhile longer, but a new model would have done better. Middleton, who was nearly eighty when he became chairman of the Commission of Revision and had done his best to improve the old model, was not the man to make radical changes. It is naturally tempting to speculate about what Bentham, with his futuristic ideas, might have done if (improbably) he rather than Middleton had guided the Commission. Confidence in Bentham's ability to apply his deceptively simple, modern principles to a very big, complex organization would be misplaced, however; the results might indeed have been disastrous (his radical reorganization of timber management was hardly an unqualified success). To have successfully worked out and applied those principles — clear lines of authority, individual responsibility, and full and accurate information — would have required a quantum leap; they still had not been realized fifty years after the Commission of Revision; indeed, it was possible for a high-powered Admiralty committee to speak disparagingly in 1885 of an eighteenth-century system adapted to the conditions of the late nineteenth century.[76]

There was no "silent revolution" between 1793 and 1815; in fact, the results of all those monumental labors were pretty meager. Still, something important did happen. For the first time a global and systematic view was taken of organization and management; a new awareness had been awakened, and the process of modernization had made an uncertain beginning.

Chapter 4

Lightening Ship, 1815-1834

Far greater and more important changes occurred in the organization and management of the dockyards between 1815 and 1834 than in the preceding period. These took place under the second Viscount Melville (1812-27, 1828-30),[1] the Duke of Clarence (1827-28)[2] and, above all, Sir James Graham (1830-34). Melville's annual visitations and extraordinarily long tenure of the Admiralty gave him an intimate knowledge of the dockyards such as very few of his predecessors possessed. Graham, however, came to the Admiralty so little informed that he put himself entirely in the hands of Sir John Barrow, the second secretary (1807-45). Barrow knew more about the dockyards than anyone else, except, of course, Rear Admiral Sir Byam Martin, the Comptroller (1816-31). Martin was fiercely defensive of the Navy Board's independence and often at odds with Melville (even though both were Tories) and especially with the Whig Graham, who retaliated by giving the Comptroller the sack and abolishing the Board.

Melville after nine years as First Lord concluded that the management of the dockyards far from having been improved by the Commission of Revision was actually no better than before. Thus in 1821 he went down to Chatham, accompanied by the junior lords and Barrow, to see for himself a dockyard actually at work. His observations afterwards were strikingly like Sandwich's fifty years before. "[Everything was left] by one Officer to the next below him, and so on downwards, nobody seeming to care or feel at all interested for the advancement of the work in hand; or for the good of the public service." The shipwrights "appeared to be working as they judged right, with little or no control from the many salaried officers appointed and paid to overlook them." The clerical officers paid "very little attention...[to] their respective duties...[and in general left everything] to be transacted by their Clerks;" indeed, it was only by questioning the clerks that Melville was able to learn how the business of these particular departments was conducted.[3]

The lethargy observed by Melville in the master shipwright's department was partly a consequence of overmanning. The 14,000 workmen employed in 1814 were at least twice as many as the yards needed in peacetime,[4] yet the workforce in 1822 numbered 10,400, the difference for the most part being the result of attrition, and the yards were on short time. The years after Waterloo were a period of deep economic depression and popular unrest; the Navy Board argued persuasively that men who had served so faithfully in wartime should afterwards be treated with compassion, while the government wanted to avoid doing anything that might add to swelling social unrest or further burden poor-relief expenditure (already at unprecedented levels). When times got better, and the unrest that had terrified the governing classes subsided, Melville at last felt free to put the higher "interest of the Public and of the Naval Service" before "the immediate (though not permanent) interest of the artificers" and in 1822 ordered the workforce to be cut over one year to 7000 (including 2500 shipwrights).[5] There were still 7700 men in 1830, however, when he ordered a further reduction to 6000. This, though, the Navy Board resisted, on the grounds that since hundreds of despairing shipwrights had already emigrated to France and Russia,[6] additional redundancies would leave the yards dangerously vulnerable if war broke out. The reductions were made anyway, and normal working resumed in 1833. Even so, the yards were still overmanned by 20 per cent (a cushion for a future emergency).[7]

The Navy Board, captivated by the notion that a time wage was a bounty for idleness, after the war persisted, with supreme illogic and predictable results, in paying a piece wage, despite massive overmanning. Indeed, rather than surrender piecework, the Board early in 1816 put the yards on short time and warned the shipwrights that their jobs would be at risk unless they slowed down. Of course, they did not slow down, whereupon the Board once again resorted to what was still called piecework with limited earnings but was actually time work with continuous check measurement.[8]

Melville, who had early on been an enthusiast for measured work, had by 1822 come round to the Commission of Enquiry's view that it had been carried too far and, indeed, was in many respects absurd, all the more so now that earnings were limited to the time rate. Yet he did nothing, fearing, as Barrow said, "that by attempting to amend it, the whole Fabric might be endangered."[9] In building, the work progressed slowly and in regular stages, in each of which it was generally uniform, well-defined and open to view and therefore easily measured. Repairs, however, were usually diverse, scattered and either ill-defined or enclosed as the work advanced; the measurers had in most cases either to make the best guess they could or simply take the shipwrights' word as to the quantity, or even the very articles, of work that should be recorded; thus errors or falsifications were almost impossible to detect. Indeed, it appears that a reasonably accurate account of about half of all shipwrights' work (by value) could be rendered only by someone constantly on the spot and would have required doubling the number of measurers then employed.[10] Such were the prices that under the system of check measurement the value of the work of all but a few shipwrights exceeded the day rate; the best men, indeed, were capable of earning 60-100 per cent more than that and often knocked off when they reached the limit (usually early in the afternoon). Everyone received the full wage anyway. At first the men resorted to intimidation against the measurers, and at one yard, at least, against the master measurer himself, when the value of their work was less than the day rate. This led in 1820 to the imposition of certain safeguards, e.g. the measurers' rounds were changed annually, and each month the master measurer spot-checked one measurer's measurements for a preceding week. Nothing really changed, though. Barrow described how by "whole sheets of Fractional Calculations, made by the Measurers Clerks, who are as numerous as themselves, every man's earnings, good or bad..., are made to come out within a penny or two pence of each other."[11]

Melville proposed to restore simple time work, but the Navy Board resisted, arguing that the shipwrights would then slack off. Graham did what the Commission of Enquiry said should have been done in the beginning and ordered a report on the relative merits of measured and unmeasured work. The shipwright officers all praised measured work but were generally of the opinion that it was inappropriate in peacetime and especially now that the yards were overmanned. Thus simple time work was restored in 1833, after forty years.[12]

The shipwrights suffered during these years such hammer blows and human degradation at the hands of the government as to leave a legacy of deep bitterness.[13]   There was great reluctance to deal with overmanning by discharging shipwrights who might at some future time be needed at short notice. Moreover, the Commission of Revision had recommended that:

> There should be so many [shipwrights], that the Officers can discharge any troublesome man without inconvenience, and that his discharge should be felt as a punishment. This will tend more than anything else to provide good order and subordination, and to enable the Officers to...[manage the yards in whatever manner] should be deemed most conducive to the Public advantage.[14]

Thus it was unskilled laborers performing work of the most menial description that were axed in 1822 — 2300 altogether (70 per cent of the total) — leaving most yards shorthanded. From this time until the late 1830's (when the workforce again began to expand) the shipwrights and other artificers at these yards were drafted on a rota system to perform the "dishonorable" work of unskilled laborers (although at their own rates of pay), to which was added the further indignity of being obliged to toil in company with the many convicts then employed at certain yards ("Felons, some of whom are hardened in guilt; and are of the worst description," was how the shipwrights described them).   Melville responded to the hapless victims' petitions by dismissing 370 men as a warning to the rest. There were no more petitions after that. Indeed, the men, in view of the depressed state of shipbuilding, had no choice but to resign themselves to their plight, however much they might detest their masters.[15]

The abolition of chip money in 1830[16] and the innovation of wage classification in 1833 were the next blows the shipwrights were obliged to suffer.   Wage classification had been recommended by both the Commission of Enquiry and the Commission of Revision as the best way of creating "an Excitement to Emulation and Exertion". The shipwrights were divided annually into three classes and paid accordingly: 4s. 6d. per day for the first class (consisting of the best men but limited to 20 per cent of the whole), 4s. for the second class (the great majority), and 3s. 6d. for the third class (men who had shown "great insufficiency, negligence or misconduct").   A gang consisted of all three classes, however, thus ending shoaling by ability adopted in 1811. The second and third classes naturally objected bitterly to classification, just, indeed, as the Navy Board had predicted they would when rejecting the Commission of Revision's recommendation years before. Graham also foresaw opposition but was not so easily deterred. It was a system, however, that stirred up poisonous jealousies and quarrels, the more so since the work being mostly collective all performed about the same amount. Yet wage classification lasted for a decade in the face of petitions and, apparently, defections from the dockyard service.[17]

Melville and Graham took very different views of supervision. Melville thought the ratio of salaried officials to workmen (it was then 1:13) was much too high. The measurers were kept busy enough; the quartermen, however, had little to do and spent much time loafing, thereby breeding indifference on their part and setting the men a bad example. Thus Melville in 1822 adapted to the dockyards the system of supervision found in commercial yards (where, however, the shipwrights were independent contractors) and dismissed the quartermen, foremen of trades, master mastmaker and master boatbuilder (the turn of the masters of the other trades came in 1830) and gave the measurers[18] the additional responsibility of inspecting the work. First-line supervision now fell entirely on the foremen of the yard; their numbers were increased (to nine at Portsmouth, eight at Plymouth and seven at Chatham), and they were relieved of clerical duties. One man in each gang[19] was now designated leading man. The leading man (another adaptation from commercial yards) received the foreman's instructions for the gang and until 1849 distributed the men's wages (for which he was paid a little extra). His functions were by no means supervisory, however; he worked along with the other shipwrights and was not salaried. The new system soon ran into difficulties. The quarterman had been a crucial, even if not always a very effective, link in the chain of supervision. That chain was now broken. It simply was not possible for a single foreman to exercise regular supervision over the hundreds of shipwrights (many working detached from their gangs) and other workmen scattered over a large dockyard;[20] in any case, the foreman usually confined his presence to just one part of the yard. Thus the master shipwright's instructions were amended in 1829, making him "almost entirely...a superintending Out Door Officer", although to no avail.[21]

Graham in 1833 created a supervisory structure even more elaborate than the one Melville abolished. It was in its essentials, however, to stand the test of time. The leading man was retained, but two gangs of shipwrights[22] constituted a company, supervised by an inspector (the equivalent of the old quarterman), and two companies made a division, under a foreman of the yard (the number of foremen was again increased). The inspector was central to the working of new system. He inspected as well as supervised his gang's work (it was at this point that measured work was discontinued) and was in constant touch with the leading men, making sure that they were at all times provided with sufficient materials and that as soon as one job was finished everything was in readiness for the next. So as to enable the yard officers to know where to give credit or lay blame for workmanship and whom to recommend for promotion or punishment, each division and company was expected to complete the work it had commenced, "from the period of first laying down a new Ship, or docking one for repair, to her completion". Thus did Graham deal with the common, and highly uneconomic but often necessary, practice of shifting men from one ship to another while the work was still in progress (now to be done only "in cases of pressing emergency"). There is, however, no way of telling how effective this new rule was; interruptions to work in progress were still frequent fifty years later.[23]

It was Graham who in 1833 created the modern bipartite division of the workforce (proposed by the Navy Board already in 1821) between permanently established and "hired" (i.e. temporarily employed) men; until then no one was entitled to, even though in practice many enjoyed, permanent employment. The rate of superannuation in 1802, and again in 1809, was increased on compassionate grounds, and to attract more shipwrights, while (a radical measure) superannuation itself in December 1814 was extended as an entitlement to all workmen, having at least twenty years unbroken service, who were discharged at the convenience of the Admiralty.[24] This, and the retention of almost the whole workforce after the war, had a double effect. First, it encouraged among the men the general opinion that no one would be discharged until he was entitled to a pension and that a pension was really deferred wages. Second, the Treasury was liable to pay over time a potentially enormous pensions bill[25] (the pensions were not contributory in principle nor were they funded, either at this time or later). Thus the Navy Board proposed creating a permanent Establishment limited to the number of workmen actually needed in peacetime; when more men were needed, they were to be hired only for one year at a time but might subsequently be eligible for vacancies on the hired list. The number (6000) to which Graham reduced the workforce in 1833 constituted the first permanent Establishment. There were a few hired men almost from the beginning, however, their numbers increasing very slowly to 461 in 1837, followed by a short burst to 1200 in 1839.[26]

Graham at the same time abolished pensions (a measure of which the Navy Board would not have approved), although not retrospectively. Already in 1828, the House of Commons Select Committee on Public Income and Expenditure had attacked pensions for workmen and recommended that they should "be immediately discontinued". Barrow advised Graham that the cost of pensions for workmen would be so great that it was certain to attract public notice and would probably be debated in Parliament but could not be defended. Nonetheless, the Admiralty soon discovered that pensions were important for obtaining good shipwrights, and they were accordingly restored (although not for unskilled laborers) in 1839. Hired time, however, did not count towards the pensions of men who were transferred to the established list; this was a deliberate omission, and the source of much future dissatisfaction. Permanent establishment and pensions were never extended either to the mills or the steam factories (the first factory opened in 1838). Indeed, there were later those at both the Admiralty and the Treasury who wished the rest of the workforce had been treated the same way.[27]

Local management acquired a more modern face and efficiency was enhanced by abolishing the offices of clerk of the survey and clerk of the cheque and transferring the business to the storekeeper. The clerk of the survey in 1822 was the first to go; he had for a number of years been under attack as in ineffective check on the storekeeper and had recently been shorn of most of his functions. The clerk of the cheque followed in 1829. The storekeeper was now the sole local accounting officer and paymaster as well; however, the appointment at the same time of a receiver of stores eased what otherwise would have been a very heavy burden, while making better provision for dealing with a highly important, but hitherto not always well-managed, area of business. Just as important, there were now two less voices with which to contend in local management.[28] The masters attendants' duties were divided many years later between the captain of the reserve and the harbormaster.

The still unresolved problem of inventory control was approached anew, first by revising store accounting in 1816, following a general survey. The procedure was each day at the close of business to check and transcribe in the storekeeper's ledger the totals in the daily transaction accounts; the daily issue accounts were verified quarterly by the storekeeper and either the master shipwright or the master attendant (depending on which department was concerned); the ledger was totalled and the accounts were closed semi-annually, the issues being carried over and deducted from the receipts column and the remains being carried forward to the next accounting period; finally, the ledger and the daily sections were checked in the Navy Office and after 1822 by the yard commissioner as well. After 1829, everyone requisitioning stores was required to give a receipt (described as "a perfect and constant check"). All of these checks existed only on paper, however, since there was no audit. A general survey at regular intervals, the Treasury pointed out, was the only effective control; otherwise, the storekeeper might easily manipulate and square the figures and never be found out. A general survey was indeed taken in 1829; twenty years passed before the next, however. Such surveys, the Admiralty argued, were just too expensive, and they continued to be taken only when the storekeeper retired.[29]

Melville attributed weak inventory control partly to the absence of any record of stores issued to the cabins and to the quartermen improperly obtaining stores from the cabins. Thus the cabins were phased out after 1822, and the foremen of the yard were required to obtain stores direct from the present-use storehouse, a procedure that put the timely provision of materials, and thus the flow of production, at risk.[30]

Responsibility simultaneously for the reception, storage and conversion of timber and the complex timber accounts (simplified again in 1825) was thought to be more than the timber master could handle. The office was abolished and the responsibilities were divided amongst several officials, either an assistant master shipwright or a foreman of the yard for reception, a timber converter for conversion, and the storekeeper for the rest. These new arrangements were a failure, however, and the timber master's department was revived in 1847.[31]

Melville, like Bentham, thought that the yard commissioner was the key to local management and ought to be endowed with absolute power, whereas his power in fact was merely nominal; he held himself responsible for nothing but acting as postman for the Navy Board and the principal officers; the officers were entirely independent of him, and no one was in a position to exercise overall coordination and control. New instructions, issued in 1822, ostensibly involved the commissioner more closely in day-to-day management by having the Navy Board correspond with him alone and having him in turn instruct the officers accordingly, convening them in the evening to consult about the following day's work and himself issuing the requisite orders (from which the officers might nonetheless deviate whenever necessary).[32]   These changes, however, were in effect procedural rather than substantive.[33]

Graham shared Melville's views on the subject of the dockyard superintendent, as the commissioner became when the Navy Board was abolished in 1832. Graham, however, thought that the superintendent's status should be higher than the commissioner's had been and also that closer naval control was necessary. The superintendent henceforth was a naval officer on active duty (a rear admiral at the principal, a captain at the secondary yards) and in the normal promotion track (dockyard service was counted the same as sea service).   Such a man, Graham argued, would naturally involve himself more authoritatively and effectually in management than the commissioner with a dead-end job had done.[34]   The principal officers continued, however, to exercise no less an ascendancy than in the past. Most superintendents were appointed either as a reward for past services or because they had demonstrated ability to command, not because they possessed managerial ability. Since they needed at least six months to learn the business, but usually served for only eighteen months or two years, most were consequently disposed to take their ease and mark time.[35]

Middleton's cumbersome Navy Board committees were finally abolished in 1829, after a long run but little applause, and their work was transferred to individual commissioners. It was at this time that the offices of Accountant General and Storekeeper General were created. Three years later, however, the Board vanished into history, and the new system had no time to prove itself.

Graham's abolition of the Navy Board, and the other inferior boards, in favor of direct Admiralty control over the civil departments has been called "the most decisive single constitutional change ever made in the administration of the Navy." Actually, it was the brainchild of Lord Grey, the Prime Minister, conceived when he himself was First Lord in 1806.[36] The measure was enmeshed in a largely political quarrel between the Whig ministry and the Tory Navy Board, and especially with Martin, who was seen off already in 1831. Graham accused the Board of financial mismanagement and of thwarting the Board of Admiralty, charges which were by no means fair but set the tone of the debate

anyway. Sir George Cockburn (for twelve years a junior lord under
Melville) did not deny that the Navy Board was "generally opposed to
any sweeping change" and "had not always...[been] ready to act as the
Admiralty wished;" nevertheless, he said, the Board had never opposed
orders when they were given, and its objections were often useful and
had saved the Admiralty from the consequences of hasty action. Graham
spoke of the need for unity and simplicity of control and of "due
responsibility"; these, he said were "the very essence and life of public
business."[37] The result, however, was a hollow unity, even greater
complexity, and a specious responsibility, while "the Admiralty...[was]
swallowed up by the duties and the attitudes of the Navy Board,"[38] and
the dockyards continued to be managed by the same methods, and no
better, than before.

The new organizational structure was an administrative nightmare and
in the case of the dockyards a disaster waiting to happen. Responsibility,
now even more than before, was diffused and divorced from authority,
and the various branches of business were worse coordinated. Each of
the five civil departments acquired by the Admiralty was administered by
a permanent official (the principal officer) — the Surveyor General,
Accountant General, Storekeeper General, Comptroller of the
Victualling, Physician of the Navy — under the superintendence of one
of the junior lords. The Surveyor was superintended by the Second Sea
Lord and was himself a naval officer; thus closer naval control seemed
to be assured. The Surveyor, Second Sea Lord and the Board of
Admiralty were all responsible for managing the dockyards and for naval
design and construction, but the Board of Admiralty alone (in certain
situations, however, one of the superintending lords) possessed
concomitant authority. The Surveyor might correspond directly with the
dockyards, but only to obtain information or explanations on
"professional" (i.e. technical) matters; he could neither issue general
orders nor "enter upon any part of the general correspondence with the
Superintendents at the Yards"; all such matters had to be submitted in
writing to the Second Sea Lord, and it was he who laid them before the
Board of Admiralty. The fact that a certain, and by no means
insignificant, part of the Surveyor's, Accountant General's and
Storekeeper General's business overlapped was a further complication,
involving a considerable additional circulation of papers and consequent
delay. The volume of extra clerical work, needless to say, was
enormous. The system, if it was to function smoothly, required that all
the parties concerned should have direct and constant access to each

other. The civil departments remained at Somerset House, however, and this, hitherto an inconvenience, now became an impediment, although it was not seen as being such at first. Communication between the Strand and Whitehall was as regular and expeditious as before; one of the superintending lords resided at Somerset House and so was always available to sign documents and superintend the more important departments. Some of the other superintending lords visited Somerset House daily, and Admiralty boards (which the principal officers might attend if the business concerned them) were held there weekly.[39] It eventually became evident, however, that the civil departments needed, in the interest of efficiency, to be removed to Whitehall, especially the Surveyor's; his was the first, although not until 1870.[40]

The composition of the Board of Admiralty added to the problem Graham had created. The principal officers were professional administrators possessing a firm grasp of the business of their departments, whereas the composition of the Admiralty Board, consisting of amateurs, was constantly shifting; there were only two Surveyors but fourteen First Lords and nineteen Second Sea Lords between 1832 and 1861. The Board was assaulted by "a relentless flow of administrative questions, many of them very detailed,"[41] which its members understood either imperfectly or not at all, a problem compounded by the disorderly manner in which the Board's business was conducted (both then and into the twentieth century).[42] Sir Spencer Robinson was moved to wonder, when he became Surveyor in 1861,

> whether establishments so vast, expenditure so enormous, details so numerous, and interests so complicated...can with prudence be committed to a governing body so variable in its composition, so short-lived in its tenure of office, and so little responsible for its acts.[43]

Graham, a great and wise advocate of financial accountability and parliamentary control, was responsible also for the Act of Parliament requiring the Admiralty annually to provide the House of Commons with an account (verified by the Audit Office) showing how the Parliamentary grants had been spent.[44] This meant little, however, since there were no means of exercising financial control; each department had its own accounts, and the Accountant General until 1885 was limited to examining these, paying bills and recording expenditure and did not scrutinize either expenditure or the estimates.[45] Graham in 1832 sent the Accountant General, John Briggs, to study the French Ministry of

Marine's accounting system in which, unlike the Admiralty's, a single
department was responsible for all accounts, the details of which were
checked locally rather than centrally. It was the opinion of Briggs that
the French system was with some exceptions superior to the British, but
Graham disagreed.[46] The Admiralty's system, nevertheless, eventually
came to resemble that of the French. Graham tried also to create a
rudimentary system for controlling the cost of labor based on the
inspectors' day books, which provided a detailed record of the kind and
amount of work that had been performed. These books were the basis
of the master shipwright's weekly expense accounts, from which the
storekeeper abstracted the annual cost of shipbuilding for the Surveyor's
information.[47] Not until 1860, however, was it possible to provide the
House of Commons with a detailed account of the cost of naval
construction.

It was Graham who in 1832 slew the School of Naval Architecture, the
dockyards' best hope for the future. The House of Commons Finance
Committee had praised the School in 1817 and Melville had stood behind
it too.[48] Now, however,

> The school was judged a failure, and those educated there were proclaimed
> in Parliament to be unfit for the higher offices of the public service. In fact
> there was no calumny with which through jealousy and official influence
> they were not assailed. The old school of Dockyard Officers sneered at
> them as amateur gentlemen shipwrights. Naval men generally disliked
> them on account of their pretensions to rank and position as gentlemen and
> men of education.[49]

Barrow, his pen dripping venom, wrote of the School's students that:

> [they had but] a smattering of science and of its practical application [and
> thus]...became a kind of mongrels and...have given...no proof of excellence
> in either. Yet these persons stand between intelligent working shipwrights
> and their Superintending Officers, whom they are to succeed excluding the
> workmen from all those situations which they formerly had to look up to;
> These apprentices are also put over the heads of all the Inferior Officers,
> and rise to the highest situations.... Happily the school has been abolished,
> but the pupils are likely to remain an incubus on the Establishment.[50]

Seven of the School's former students left the service almost at once
in despair, while those who remained sank ever deeper into

disillusionment. In 1838, twenty years after the first students had passed out, not one had been promoted to master shipwright, and only one was an assistant master shipwright.[51] Eventually, of course, they did inevitably come into their own; Isaac Watts, the Chief Constructor of the Navy, who with Scott Russell designed the first British ironclad (the *Warrior*, 1860), Watt's two assistants, and all the master shipwrights of that day had attended the School of Naval Architecture.

The changes decreed by Melville and Graham were of vastly greater scope than those in any previous period and reflected the rising standard of public service. Were the dockyards better managed as a result, however? In some respects they were, but in general they were not. More than twenty years of heavy overmanning and, at the same time, the use of check measurement would not be looked upon today as evidence of good management. Everyone — officers, supervisors and workmen — became habituated to underutilization of labor. Shipwrights and other highly skilled workers were demoralized and industrial relations were poisoned by check measurement, compulsory performance of common laborers' work in association with convicts, the abolition of chip money, and wage classification. Graham's abolition of the Navy Board was certainly a bold move; however, he substituted a committee of amateurs for a committee of professionals, created a disjointed and unwieldy management structure, withheld managerial authority from the permanent official (the Surveyor) responsible in fact for managing the dockyards, and perpetuated the centralized system of management by correspondence inherited from the eighteenth century. Melville and Graham, like others before, thought the yard commissioner (or superintendent) should play a large role in local management; his role continued, nevertheless, to be minimal in practice. Melville's simplification of supervision was a mistake which, however, Graham corrected; indeed, Graham's system was superior to the one Melville abolished. Melville streamlined local management, eliminating the clerk of the survey and the clerk of the cheque and consolidating accounting under the storekeeper. His elimination of the timber master and restructuring of timber management was another mistake, yet was driven by the same logic, and by the misperception also that the practical and the accounting sides of the office were incompatible. Procedures for requisitioning materiel were simplified and tightened up; eliminating the present-use cabins, however, was yet another mistake (only rectified years later) and had an adverse impact on the flow of production. Melville tried also to improve store accounting and inventory control. Overall, certain important long-term organizational efficiencies were achieved, but in the context of the inefficient eighteenth-century system of organization and management ratified by the Commission of Naval Revision.

Chapter 5

Winds of Change, 1834-1854

The structural and systemic deficiencies hindering the management of the dockyards still had not been dealt with as the nineteenth century approached the mid-point. Moreover, the steam revolution, only recently begun, had resulted in a new department and new problems. It was, however, the enormous cost of the race with France to build a steam navy and the requisite support facilities (dockyard expenditure doubled between 1837 and 1849) that once again drew attention of the dockyards and led between 1846 and 1849 to a series of inquiries into how well they were managed. These exposed troubling weaknesses and were followed by another spate of reform. At the same time, Graham's reorganization of the civil administration (although not the principle of management by correspondence) came under fire also. The apprentice schools were firmly established; the School of Naval Architecture was very briefly restored; and the merit system for promotions was introduced.

The steam navy began with a few tugs in the 1820s and by 1837 consisted of twenty-seven paddlewheel vessels, including two frigates (the first) under construction at Pembroke. The Admiralty was never to build its own engines; it was self-sufficient, though, in installation and repair facilities. A chief engineer was appointed in 1835; two years later the steam department (under the Comptroller of Steam Machinery) was created; and in 1838 the first steam factory came on line at Woolwich yard.[1]

The French soon after embarked on an ambitious program of modernization, and the Admiralty responded in 1842 by laying down twenty-nine steam warships, reviving piecework, and expanding the workforce. It was at this time that Deptford yard, closed in 1832, was reopened and provided with new building slips and that a second steam factory was built at Portsmouth. A third factory, with fitting-out basin and docks, was subsequently provided at Keyham, adjacent to Devonport yard (as Plymouth yard was now called); however, the work was stretched out over a number of years and the factory did not become fully operational until the Crimean War (when the factory and the yard were connected by tunnel). Meanwhile, the problems of screw propulsion had been worked out, making steam-powered line-of-battle ships practicable, and there were five such vessels (all but one converted from sail) by 1852.[2] The steam navy was controversial, however, and its creation went forward very slowly until the Crimean War.

"The sums expended in the Dockyards seem to be so totally disproportionate to the work done," the First Lord, the Earl of Ellenborough, said in 1846, "that...there must be a want of due supervision and control, and a great defect of system." His successor the Earl of Auckland, concurred and sent R.S. Bromley, an Admiralty official and later Accountant General of the Navy (1854-63), to visit the yards. Bromley's confidential report (February 1847)[3] was followed by Auckland's admonition to the superintendents that "the quantity of work is below the standard of well-conducted private establishments, and that there is great room for improvement in the application both of stores and labour." A year later a House of Commons Select Committee expressed similar doubts and recommended a full inquiry into the management of labor and wages, "the cost of the work, the reception of stores [especially timber]..., and...the method of keeping the accounts", whereupon Auckland appointed (October 1848) the Committee of Dockyard Revision consisting, most notably, of Bromley, Captain Baldwin Walker (the new Surveyor of the Navy) and Henry Ward (the Admiralty secretary). The committee spent eight weeks visiting the yards and reported in December 1848. Auckland died quite suddenly on 1 January 1849, however, and the report was implemented by his successor, Sir Francis Baring (1849-52), who had been a member of the Select Committee.[4]

The great problem was how to integrate the steam factories into the management structure of the dockyards. They constituted at first a separate Admiralty department under a principal officer, the Comptroller of Steam Machinery (a naval officer), and each factory was managed by a chief engineer coequal with the master shipwright. Both chief engineers were well versed in marine architecture and ship construction; Thomas Lloyd (Woolwich) had been educated at the School of Naval Architecture before taking up marine engineering, and Andrew Murray (Portsmouth) was a marine engineer and former managing partner in William Fairbairn's Millwall yard.[5] The master shipwrights, however, had very little knowledge, and no experience, of marine engineering; they disdained the engineers, regarding them as dangerous "upstarts" and rivals and generally refused to cooperate, or even in some cases to communicate with them. There was much friction. The two departments often worked at cross-purposes, the master shipwright hurrying the completion of one ship, the chief engineer at the same time working flat out on the engines for another, and neither knowing what the other was doing. Indeed, the Surveyor did not even provide the engineer with the plans of the ships that were to be fitted with boilers and machinery. Thus the hitherto independent steam department in 1849 became a branch of the Surveyor's department (under Lloyd, as Engineer-in-Chief), a solution which may have been logical but simply did not work, since the chief engineer and the master shipwright remained independent and coequal officials.[6]

The failure to coordinate was by no means a failure of the construction and engineering departments alone. Rigid Admiralty control over the dockyard officers, and the no less rigid organizational structure of the Admiralty, vitiated individual responsibility and undermined interdepartmental coordination, as, indeed, Bromley (going beyond his instructions) pointed out in his report to Auckland. He would probably have agreed with the anonymous pamphleteer who argued at this same time that it was a system that could *"never stand long in the case of war, nor work well in time of peace."* Bromley thought that since the management decisions of the Board of Admiralty and the superintending lords were taken on the advice of the principal officers, authority and responsibility should be delegated accordingly. Anticipating the objection that "this would place in the hands of a subordinate officer the power of virtually overruling the decisions of his principals," Bromley recommended, albeit unsuccessfully, that the principal officers should meet weekly as a committee chaired by one of the junior lords "for the purpose of discussing all matters arising out of their duties beyond their defined powers as separate officers."[7]

The Surveyor's department was reorganized, responsibility for ship design and for the management of the dockyards and the building program in effect being separated, much to the advantage of the latter. It was Graham's intention that the Surveyor should be a naval architect as well as a naval officer; for this reason he chose Captain William Symonds, whose undoubted distinction as a naval architect appears, however, to have overshadowed his abilities as a manager. Auckland was led by the evidence to conclude that the two jobs were more than one man could handle without slighting either the one or the other. Thus when Symonds retired in 1848, he was succeeded by Walker, a naval officer possessing "sound practical knowledge and ability as a Seaman rather than...[expertise] as a Ship Builder." Henceforth, management of the dockyards was the Surveyor's particular responsibility, and ship design was entrusted to the two assistant surveyors (the second being appointed at this time), the Surveyor confining himself to submitting their plans, and his own remarks thereon, to the Board of Admiralty.[8]

The Committee of Revision found no evidence of "grave...misconduct, or peculations" having occurred in the dockyards themselves in the preceding twenty years; indeed, the officials in "many" departments were praised as being able, upright and hardworking men. What it did find, however, was pervasive slackness from top to bottom: "discipline has been relaxed, — additional labor has been employed to do that, which additional industry might have accomplished." So little had the dockyards changed since the eighteenth century. Yet the Committee, like Bentham, refused to blame individuals; it was rather the system of organization and management that was at fault. "There is a want of unity about it, and of real responsibility, which generate laxity, and must lead, ultimately to abuse."[9]

The workforce nearly doubled, from 6000 to 11,722, between 1833 and 1848. The number of established men grew little by little to 9452 as exceptions were made, even though, contrary to the late Navy Board's fears, it was generally easy to find quite enough men (including shipwrights) who would accept temporary status. Indeed, hired men were under the impression that their jobs were permanent anyway, partly because of a natural reluctance to dismiss anyone when the work he had been hired to do was finished and partly because political pressure made it difficult to sack large numbers of men who were likely to join the ranks of the unemployed. The conclusion drawn by the Committee of Revision was that the number of workmen had increased rather more than the quantity of work between 1843 and 1848. The workforce was consequently reduced to 9621 men (all established), the number considered sufficient to fulfill the current naval program.[10]

The master shipwrights revived piecework in 1842. Bromley, however, condemned it, for the same reasons as the Commission of Naval Revision. Ascertaining the quantity of work actually performed, and even identifying the articles themselves, was, he said, "almost...guess work" (the inspectors measured the work); large numbers of articles were mispriced; some men were paid as much as double the day rate (then 4s.) while others were paid less than they should have been and thus were discontent. The master shipwrights were naturally aware of these problems but defended piecework anyway, arguing that it increased productivity, as, indeed, it probably did to some extent, but only because of the utterly demoralized state of the men following their victimization in the 1820s and 1830s (it was only in 1842 that wage classification was abandoned). Nonetheless, piecework was discontinued in 1847, largely on the grounds that the expense could not be justified in peacetime.[11]

It was at this time that the master shipwrights prevailed on the Admiralty to introduce occasional check measurement, for which only twenty-two measurers were needed, since just the work of gangs suspected of ca'canny was measured[12] — first the work of the previous week and then, for comparison, that of a subsequent week. The Committee of Revision endorsed "back measurement" (as the men called it) as being a highly valuable adjunct to supervision. In the opinion of the superintendent of Portsmouth yard, Rear Admiral Sir Willliam Fanshawe Martin,[13] a decade later, however, it was "worse than useless"; the men having been warned by the first measurement naturally increased their exertions meanwhile, but only for the time being. In any case, wages as a rule were not stopped if the men complained enough, while the supervisor who was foolish enough to report a man might well become the butt of harassment in the shape of graffiti or scurrilous letters in the post, or, if he had made himself sufficiently odious, might be beaten up by the offended workman and his mates.[14]

Auckland attributed weak work-discipline, in part, to the inspectors' heavy clerical duties (i.e. making out requisitions), which interfered with supervision, and to a lack of incentive on the part either of the inspectors, whose salaries were so small (£100 per annum, or less than a leading man, who shared piece earnings, might make), or of the leading men, whose prospects of being promoted to inspector were quite slender.　　Thus in 1847 two classes of inspector were created, a supervisory first class (paid £150 per annum) and a clerical second class (paid £125), and the number of inspectors was increased.[15]

Auckland looked for a substitute for "that personal, and pecuniary, interest [i.e. the profit motive], which is the mainspring of exertion in private Establishments," and thought it was to be found in suppressing "the Nefarious traffic in dockyard appointments" which created the feeling that promotion from workman to leading man to inspector was "a matter of accident or favor". Every man should know unequivocally, he said, "that his conduct will be known, and appreciated, by his Superiors, and that, however humble his position originally, his future fate depends upon his own exertions." Thus was the prototype of the modern merit system introduced in 1847. Each workman and supervisor was evaluated by his superiors (they were admonished not to let their reports "lapse into a matter of form"). When a position became vacant, the evaluations of each eligible man were reviewed by the master shipwright, and three candidates were nominated; the superintendent vetted the nominees (a precaution against improprieties); the nominees were examined in the practical and theoretical aspects of their prospective duties; the names of the two judged to be best qualified were forwarded to the Surveyor, and he made the final choice, subject to concurrence by the Board of Admiralty.　　The House of Commons Select Committee on Dockyard Appointments in 1852 declared itself satisfied that the system had served its purpose well; the superintendents expressed a "belief in the honesty and good faith of their principal officers", and Auckland's successor, Baring, observed that supervision had improved to such a degree that the same amount of work was done as before, even though the workforce was much smaller. Still, the merit system was riddled with ambiguities and contradictions.　　The examinations were prepared and evaluated locally and in practice were given only when all the persons who had taken the previous examination and been recommended had been promoted. Moreover, the superintendents were not bound by the results and might instead base their recommendations on the personal evaluations made by the candidates' immediate superiors.　　It is therefore not surprising that suspicion that the right connections often counted more than merit in making appointments continued to be voiced.[16] Nonetheless, the merit system, despite its flaws, was a big step forward.

The dockyards were a paradigm of massive immobility. Nothing makes this clearer than the management of timber. The Committee of Revision was staggered by the sheer disorder it observed and by the primitive methods employed in shifting heavy materials. Except at Chatham and Pembroke, the "want of Method, and absence of proper economical arrangements" for the reception, storage and preservation of timber was "scandalous"; thickstuff and plank were not properly classified and stacked, nor were the quantities duly recorded; the "waste of time, money, & material" was as a result "incalculable". This may in part have been a consequence of the timber master's department having been suppressed, and the business dispersed among various other officials, in 1827. There were, it is true, extenuating circumstances at Portsmouth and Devonport, since those yards were somewhat disrupted by the construction of the steam factories and related facilities. Nevertheless, not a single yard had been provided with a place for surveying timber near the landing wharf, despite unusually heavy deliveries (for many years now) because of the naval program. Nor had such measures been "taken for its subsequent removal, as would have suggested themselves...to the minds of the most ordinary contractor". In several yards, four horses and three men were observed shifting a piece of timber which, had a traveller been used to place it on a truck, could have been moved by one horse and one man. Moreover, timber might be moved six times before being deposited in its final resting place and usually so far from the saw pits that it had to be moved yet again. Indeed, at the western yards "the clumsy contrivances" and brute labor employed in moving heavy materials contrasted starkly with the railways or travellers and trucks used by the contractors.[17] The timber master's department was revived already in 1847, yet the Committee on Dockyard Economy in 1859 was no less critical of timber management than the Committee of Dockyard Revision in 1848.[18]

These "slovenly, & costly, blunders" the Committee of Revision explained in language long since familiar: no one considered himself responsible or really cared.

The Timber Inspector says he cannot act without the concurrence of the Master Shipwright; the Master Shipwright considers, and not without reason, that, in the general arrangement of the Yard, he ought to have the sanction of the Superintendent; — while the Superintendent takes no general view of the economical wants of the Dock Yard, and leaves the details of the timber arrangements to his subordinates.[19]

Sandwich, Middleton and the Commission of Naval Revision had all emphasized the centrality of the dockyard regulations to good management. The Committee of Revision observed, however, that while the regulations were indeed "admirable", their efficacy depended more upon the spirit in which they were carried out "as a whole" than upon "rigid adherence...to particular parts". This was a novel, and highly important, insight, even if it did not bear any fruit.[20]

Each dockyard had to "be dealt with as a whole", the Committee said; it was like "a machine, the working of which depends upon the Systematic Action of all its parts; — a great manufacturing Establishment". Bromley pointed out in his report, however, that "at almost all the yards, with the exception of Pembroke, there is an absence of that cordial communication between the superior officers, and cooperation in their duties, which the interest of the public service demands." The Committee observed that the yards were not ruled and animated by that vigilant self-interest which made for good management in "Private Establishments of a similar kind". The dockyard superintendent was the only safeguard against this and thus must be "a substitute for the Master's eye." He appeared from his instructions to be "everything" yet in practice was nothing, "the mere organ of the Permanent Officers". The Committee recommended choosing him for his managerial qualifications, having him serve long enough to make it worthwhile for him to master the job (advice which was not taken), and identifying him more closely with the actual management of the yard. Thus a quarterly board of estimate consisting of the superintendent (chairman) and the principal officers, was created in 1849. The function of the board was to review "the progress, and requirements, of the yard," so as to establish more effective control over operating costs, although in practice it was chiefly concerned with stores. The officers had little interest in making the board succeed, however, and in fact it did not succeed.[21]

Stores were very loosely managed. Inventory had not been taken since 1828; thus it was not possible to rely on the accuracy of the storekeeper's ledgers. "No Private Establishment would go on for...20, years, without taking stock," said the Committee of Revision. "No Public

Establishment can inspire, or deserve, Public confidence, without subjecting itself to a similar ordeal." The quantities of stores requisitioned quarterly by the officers appeared to be excessive. Indeed, it often happened that a yard would spend more money for stores than Parliament had voted. Several corrective measures were adopted on the Committee's recommendation. The store establishment was updated and triennial surveys (including timber, which had never yet been surveyed) were instituted beginning in 1849.[22] Responsibility for ascertaining the yard's requirements was transferred from the storekeeper (who usually accepted the officers' requests at face value) and the Storekeeper General jointly to the board of estimate. The board vetted the officers' and the storekeeper's requests, and the superintendent certified to the Storekeeper General that "the legitimate requirements of the yard" had not been exceeded.[23]

Although new technology and keener competition were pushing the prices of many items down, the standing contracts had nevertheless been permitted at great cost to run on for as long as twenty or thirty years. Indeed, when the system was revised and new bids were taken in 1848, the Admiralty in many cases benefited handsomely (by as much as 36 per cent), leading the Committee of Revision to recommend that contracts should be reopened every three or four years.[24]

The realization that accounts, if they were properly structured and carefully compiled, could be a useful management tool was one of the most important developments in this period. Here Bromley, influenced by the example of private industry, appears to have played a central role. Accounts there were aplenty, but they were mostly based on unsound principles and were unreliable and extremely complicated (only one or two clerks at each yard understood the timber accounts). The Surveyor received each yard's annual expense accounts (one for ship construction, another for yard services), but no use was made of these and consequently little care was taken in compiling them. Bromley told Auckland that these potentially vital accounts required

immediate revision [so as]...to lay [in the Accountant General's department] the foundation of a more general and useful account, by which [following the practice of large manufacturing firms] the entire expense of a naval arsenal might be ascertained and the progressive cost of every ship, work, manufactory, and establishment contained in it, might be recorded from month to month at the Admiralty.

He had no doubt that it would be a tremendously formidable but not, as he was sure some would say, an impossible task. In the event, however, the order for monthly expense accounts (1849) was observed only very briefly; nonetheless, Bromley was able to revive these accounts when he became Accountant General in 1854.[25]

The accounts of yard wages were never checked for accuracy (in contrast to the meticulous, although ineffectual, checks on store issues), and the men were often overpaid as a result. The yard provided the Accountant General with weekly abstracts of wages, and from these he prepared for the Board of Admiralty his own statement and indicated the proportion of each yard's wages to the money voted by Parliament; he never saw the original, detailed records, however. The Committee of Revision rejected centralized control by the Accountant General[26] (this would have necessitated transporting "a mass of Vouchers, weekly, to Somerset House" and would have led to endless inquiries and correspondence) in favor of local control. A store clerk was instead assigned to the superintendent's office to check the weekly wage accounts and prepare them for local audit. As a further check, the Accountant General (or his deputy) was directed to visit the yards and inspect the accounts annually.[27]

No one could say with any degree of accuracy how much labor had been expended on a particular ship. The inspectors' day books, forming the basis of the abstract of shipbuilding labor (compiled annually for the Surveyor), were almost "universally inaccurate." The Committee of Revision's solution — instructing the master shipwrights and foremen to exercise greater care in examining the books — was, of course, no solution at all. Bromley in 1856 succeeded, however, in having the first dockyard accountants appointed, and the functions of the old pay office were transferred to these local agents of the Accountant General.[28]

The contrast between conditions at the Portsmouth and Woolwich steam factories provided vivid evidence of the connection between efficient accounts and sound management. The Committee of Revision had nothing but praise for Portsmouth's factory (even though it was not yet fully operational) and the accounting procedures which Andrew Murray, the chief engineer, had brought there from Millwall. "There was not a man, out of the 533 men employed, whose work on any particular day..., could not be at once ascertained, & accounted for." Moreover, Murray was planning to take inventory quarterly. The Committee had little good to say, however, about the Woolwich factory; there it was dismayed to find "a want, throughout, of proper method, and responsibility." A "very ingenious system of accounts" had been devised, but the crucial account of labor was based on records compiled on slates by the workmen themselves, and inventory had never yet been taken. "When the basis is rotten," said the Committee, "we place little faith in the superstructure." Not all was by any means the fault of Thomas Lloyd, the chief engineer, however. Lloyd was obliged to start from scratch; the Admiralty formulated no general instructions on how the factory was to be run, and Lloyd did not have Murray's advantage of having been in business; the factory had grown too fast (it employed twice as many men as Portsmouth) and had been "overloaded with work"; and the officials had often been "detached for other services." The Comptroller of Steam Machinery was prodded into drawing up general factory regulations, and the Committee of Revision and the two chief engineers together revised and assimilated the Portsmouth and Woolwich accounts to each other.[29]

The part-time apprentice schools proposed by Bentham and the Commission of Naval Revision were established between 1843 and 1846 and were soon providing the short-lived second School of Naval Architecture (1848-53) with students. The schools, however, had other functions as well — to shape a more efficient workforce by inculcating habits of discipline and subordination and by imparting useful knowledge, and to provide better educated and more effective supervisors. Thus were they a precocious embodiment of human-resources management ahead of anything in the private sector.

The closing of the old School of Naval Architecture was seen to have been a mistake; unless the corps of constructors whom it had educated (and who were conducting a vigorous campaign on their own behalf) was replenished, the future of British naval power, now entering the age of steam in competition with the French (whose own school of naval architecture was now a century old), was likely to be at risk. In order to avoid the (unfounded) criticism that had killed the old school, the new school only accepted students from the apprentice schools. Apprentices who aspired to responsible positions were already attending private part-time schools kept for their benefit. Many parents, however, could not afford the tuition or even provide their sons with any education at all; indeed, the high proportion of illiterate apprentices, quite apart from any other considerations, made establishment of the apprentice schools imperative. Attendance was compulsory for the first three years, and a few bright boys were selected to study an additional year during which they were instructed in mathematics, chemistry, mechanics and the principles of naval construction. Some of these lads were then admitted by competitive examination to the School of Naval Architecture and became, after four years, naval constructors, while the most promising of the rest went to the mould loft and were subsequently appointed either draftsmen or yard inspectors. The School of Naval Architecture, however, was soon said to be educating more constructors than were needed and was abolished, ironically in the same year that the Science and Art Department of the Privy Council was created and only two years after the Great Exhibition. Not until 1864 was there to be another school of naval architecture. Meanwhile, the education of future constructors consisted of the little the apprentice schools had to offer.[30]

The apprentice schools were a unique recognition of the value of technical education in a country where the attitude of industrialists was almost universally apathetic, or even hostile. Still, there was at first little to inspire confidence in the schools. It was intended that education should be a factor in appointing apprentices, but there was no examination and, in practice, patronage was often more important, e.g. the Admiralty in 1846 appointed only one of the four most meritorious nominees on the Portsmouth superintendent's list but five "from a class distinguished from the rest by the schoolmaster as 'very ignorant.'" An entrance examination (part of the new merit system) was instituted in 1847 but being confined to the three Rs was not very formidable; however, algebra, geometry, grammar, English composition and geography were subsequently added. The examination was made more difficult in 1860; henceforth, it was administered by the civil service commissioners and apprentices were appointed strictly on the basis of placement; as a result, the number of candidates fell and the scores rose markedly. The educational environment was further improved by reorganizing the schools in two divisions — an upper school for the most promising boys and a lower school for the rest.[31]

The few years from 1846 to 1849 were enormously important not so much for what was actually accomplished as for new insights which were to deepen and lead eventually to radical change. Many but, significantly, far from all, of the yard officials were found to be able, upright and hardworking, even though there was little to show for it. Output per man was less than in commercial yards; the construction and engineering departments were uncoordinated; timber and stores were ill-managed;

accounts were all but useless; no one could say how much labor or materials were actually expended on any particular ship; crippling slackness was still pervasive. The organizational structure adopted by Graham for the Admiralty was seen by some to have weakened management. The Committee of Revision said that each yard should be treated as a unit, not as a collection of separate parts, but the board of estimate brought the officers together only occasionally. Putting the steam department under the Surveyor did nothing to help local coordination, and measures to involve the superintendent fell far short of making him a general manager. It was seen that literal observance of regulations did not lead to good management but not that rigid Admiralty control stifled initiative and encouraged slackness. The dawning perception of accounting as an instrument of management was a highly significant development, as was recognition of the apparently superior managerial ability of engineers over master shipwrights. The merit system was at least an ingenious, even if it was unlikely to prove to be an adequate, substitute for the engine of profit that drove private enterprise. The system in tandem with the apprentice schools promised, however, to produce better educated and more intelligent and efficient supervisors and workmen. On the other hand, abolishing the second School of Naval Architecture was a retrograde step. Nevertheless, the management of the dockyards by 1849 had been set, even if uncertainly, on a modernizing course.

Chapter 6

Under Siege, 1854-1868

The steam revolution had barely begun when the Crimean War broke out; the war itself was barely over and the iron revolution began. The ships of the new naval age were much bigger and cost much more to build than the old sailing ships which they now rapidly displaced. The steam factories had been built, existing docks had been lengthened and new docks built, all at enormous expense. The cost of yard maintenance, which stood at less than £100,000 in 1851, averaged £570,503 per annum between 1853 and 1858, while the workforce between 1852 and 1858 grew from 9960 to 12,215 in the dockyards and from 1046 to 2361 in the steam factories. Yet the Surveyor, Sir Baldwin Walker, was saying in 1858 that the dockyards were inadequate and in the event of a long war could not maintain a fleet capable of protecting Britain's own coasts. The Navy's first ironclad warship, the *Warrior*, had to be laid down in a commercial yard; none of the royal yards was equipped to build her. The dockyards were by now a byword for inefficiency and at risk, as major building yards, from the economic doctrine of the age. No longer were efficient financial and cost controls, an efficient organizational system, and reliable information systems merely desirable; they were imperative.[1]

The French in 1858 launched a vast naval program to expand their steam navy to 150 ships and began enlarging their dockyards accordingly. This was a challenge to British naval superiority which could not be ignored. Moreover, four of the French frigates were to be ironclad, and the first of these, the *Gloire*, was launched in 1859. This even greater challenge the Admiralty answered with the *Warrior*, built by the Thames Ironworks and launched in 1860. The *Warrior's* hull was armored iron, a more important innovation than the *Gloire's* armor-clad wooden hull; the *Warrior* was also much bigger than the *Gloire* and in most other respects superior. Thus began the race between guns and armor that was to dominate warship building down to 1914. Still, the French had learned better than the British the lesson of the Crimean War that wooden walls were useless against shell-fire and had concluded that wooden ships would be obsolete in a decade, whereas the Admiralty, characteristically taking a more cautious view, laid down several more wooden ships of the line, and even increased the timber establishment, while building the *Warrior*.[2]

The impression that the dockyards were inefficient and extravagant gained ground, and the First Lord, Sir John Pakington, acting on the instructions of the Cabinet, in 1858 appointed the Committee on Dockyard Economy. The Committee's report in 1859 was a swingeing indictment of the system of local management based not on the evidence given by the dockyard officers, who, on the contrary, thought the system was basically sound and only needed to be fine-tuned, but on "the practice of private trade."

> Errors and faults will gather and increase with the age of any establishment, and few establishments of such old standing as the dockyards, will bear comparison with private establishments; but it is this comparison which the Committee have endeavoured to bring to bear upon them....[3]

These strictures should not have been very surprising considering the Committee's composition. Only one of the five members — Henry Chatfield, the master shipwright at Deptford — was a true insider. Two were engineers and either had been or were businessmen — Andrew Murray, the chief engineer at Portsmouth, and Robert Bowman, a civil engineer and shipbuilder. Robert Laws, the storekeeper at Chatham, represented the accounting side and was strongly influenced by commercial practice, as was Rear Admiral Robert Smart, the Committee's chairman. Chatfield, the odd man, refused to sign the report and contended in a vigorous dissent that the dockyards were "conducted upon a system in every way worthy of their object" and according to regulations "based upon long experience"; successive naval administrations had paid too much attention to the dockyards for them to have fallen "into such a state as that represented by the Report of the Committee;" on the contrary, management had steadily been improving. All the more stinging was the omission from the report of any similar strictures on the steam factories. Moreover, Chatfield, like all the shipwright officers (and naval officers too), despised the upstart engineers and their new-fangled notions of efficiency.[4]

At just this moment, the government changed hands, Lord Palmerston and the Liberals succeeding the Earl of Derby and the Conservatives. The new ministry was caught at once in an erupting controversy rivalling the one fomented by St. Vincent and the Commission of Naval Enquiry at the beginning of the century.

Observations on the Committee's report, from the Admiralty and the yards, swelled to the proportions of a flood. Walker, Bromley, the dockyard superintendents and the master shipwrights all took Chatfield's side, whereas the storekeepers and recently appointed yard accountants were more receptive to the Committee's views. Finally, the First Lord, the Duke of Somerset, rejected the report's "expression of indiscriminate blame" and chided the Committee for having ignored the Surveyor's department; it was there, he said, that its inquiry should have commenced.[5]

A further inquiry was inevitable, and the government accordingly appointed the Royal Commission on the Control and Management of Naval Yards (1860-61). The Royal Commission agreed with the Committee on Dockyard Economy and concluded, in a succinct and lucid report, that the dockyards indeed were not well managed; the structure of the Board of Admiralty and the organization of the subordinate departments were defective, and there were neither "clear and well-defined [lines of] responsibility" nor "any means...of effectively controlling expenditure, from the want of accurate accounts."[6] A finger had been put authoritatively on fundamental problems that somehow had to be worked out if the dockyards were to be managed with the same results as the commercial enterprises with which they were so disparagingly compared.

The general opinion was that armored ships should be built in commercial yards, as the *Warrior* had been. That of course was not very desirable from the Admiralty's point of view, nor was it even possible, in view of the vastness of Somerset's naval program. For the Admiralty between 1862 and 1866 built armored iron and composite ships[7] and converted wooden ships-of-the-line to ironclads in an all-out, and successful, effort to overtake the French, whose own program, however, remained largely on paper. The construction of armored ships was nevertheless restricted to Chatham and Pembroke. The two yards were quickly re-equipped and the first dockyard-built iron warship (by specially hired platers), the *Achilles*, was launched at Chatham in 1863. Repair and maintenance needed to be performed at Portsmouth and Devonport, however, and that necessitated furnishing those yards with similar equipment, although to a lesser extent. Soon they too were building iron ships, for repair work was irregular and it did not make economic sense to leave so much fixed capital idle for long periods. It was soon discovered, moreover, that iron ships needed to be repaired less often than either wooden or composite ships. Iron ships, however, were so much bigger than the old wooden ships that vast sums had to be spent on either lengthening existing, or building new, slips and docks and on enlarging the fitting-out basins. A dock was enlarged at Portsmouth already in 1861, so as to be able to receive the *Warrior*, and Parliament in 1865 sanctioned the expenditure of £2.5 million for the extension of Portsmouth and Chatham yards.[8] All of this expenditure, and more brickbats,[9] ensured that the dockyards remained an object of parliamentary interest.

The triumph of iron over wood had other repercussions as well. The very considerable stocks of large timber on hand in 1861 had been used up by 1868, by building composite ships; iron ships nevertheless contained much small timber. Thus the timber converter survived but not the timber master, whose duties were now divided between the timber converter and the master shipwright.[10] More important, the workforce was inevitably restructured, although on lines quite different from those in private yards.

Meanwhile, Walker retired (he had been in office for twelve difficult years, for the last months with the new title of Controller of the Navy), and was succeeded by Rear Admiral Robert Spencer Robinson[11] at the beginning of 1861. The choice of Robinson proved to be as happy as it was natural. As a member of the Royal Commission, he stood head and shoulders above his colleagues, and compiled a highly perceptive and invaluable digest of the evidence (appended to the *Report*). Later he proposed to Somerset, and then laid aside, a brilliant, but at the time much too radical, scheme for reorganizing the Admiralty departments and local management. The scheme, however, formed the basis of the ill-fated reforms carried out by H.C.E. Childers after 1868 and will be discussed in that context in the next chapter. As Controller, Robinson attacked the management problems of the beleaguered dockyards with intelligence, imagination and energy; he was indeed the best, and the most cerebral, person ever to hold that office.

The management of the dockyards did not break down during the Crimean War as had been predicted, although Walker was certain that it would do so in the next war. Responsibility, authority and control were all hopelessly divided at the Admiralty, he told the Royal Commission; the principal officers possessed none of the authority and responsibility without which it was impossible to function effectively as managers; they might issue no orders without the approval of their respective superintending lords and, through them, the Board of Admiralty; nor might an order be altered without approval. Walker was "held responsible for the works going on [in the yards] without having an adequate power to control them". So cumbersome was the system that it was often impossible in wartime (or at any time when the yards were working flat out) for him to communicate orders in a timely manner, even in cases of the greatest urgency; it was only because he circumvented the system and issued orders entirely on his own authority that the system had not collapsed during the Crimean War. So false, indeed, had he felt his position to be that only the urgent entreaties of Sir Charles Wood, the First lord, dissuaded him from resigning in 1857. Somerset dealt with the problem already in 1860; Walker was made responsible for dockyard work and expenditure on labor and empowered to give orders directly to the superintendents (and was for this reason given the title of Controller of the Navy).[12]

Walker complained of having to go through the Storekeeper General's superintending lord in all matters relating to materiel, of precious time being lost shuffling papers between ill-coordinated departments, of contradictory orders being issued, of correspondence of vital importance being misdirected and never reaching the Controller. The officers at Devonport had at one point written no less than 114 letters to the Storekeeper General pointing out that timber shortages were hampering production, yet only ten of the letters were forwarded to Walker. How, he asked, could the yards be managed in a satisfactory manner when the Controller was not kept properly informed of such matters?[13]

The Royal Commission's recommendations for establishing the sine qua non of "clear and well-defined responsibility" were not carried out at the time; they were far too radical. The First Lord, the Royal Commission said in its report, should be a minister of marine;[14] he should be entirely responsible for the control and management of the dockyards, and the principal officers should be appointed by, and be responsible to, him. "The dockyards should be looked upon as large manufacturing establishments"; the Controller should be "acquainted with and qualified to manage such establishments" and should be empowered to choose the superintendents (subject to the First Lord's approval), since they were "the instruments for carrying out his instructions." Likewise, the superintendents should have the same power of choosing their own subordinates. The Storekeeper General should be under the Controller; Robinson had insisted on the importance of this, arguing that the Controller should at all times be fully informed about the expenditure of materiel.[15]

It was of the utmost importance, the report went on to say, that "all the departments...connected with the dockyards should be brought under one roof." The Controller's department since the late war had been housed partly in Whitehall and partly in Charing Cross (a division which itself created problems), but the other departments were still at Somerset House. Thus officials wasted valuable time committing to paper what they could have explained in person in a matter of minutes, the double-recording and duplication of correspondence had swelled to enormous proportions, and 107 messengers were needed to shuttle papers back and forth. The Treasury in 1862 gave approval for an addition to Admiralty House big enough to house all the departments there; another twelve years passed, however, before the new quarters were ready.[16]

The Dockyard Committee took the view that management was the key to productivity. Less work was done in the dockyards than in well-run private shipyards, not because the dockyardmen were less "ready to do a fair day's work for a fair remuneration...[than] any other men," but because many of the master shipwrights and their subordinates were "incompetent, indifferent or careless" and there was no "mutual confidence"[17] between management and labor.

The Committee deplored the elaborate substitutes — centralized decision-making, minute regulations and coercive measures like check measurement — that had been adopted to compensate for managerial and supervisory shortcomings but only compounded the problem. "[No] code of rules for the management of such large bodies of [work]men...[could ever] enable an inefficient or inactive professional officer at the head of a department to carry it out properly." Naval officers were useless as superintendents; even "the most active naval superintendents" had failed to accomplish very much; the officers would be far more effective if both they and their subordinates were able and active men.   Careful observance of the morning meeting was essential (at some yards it was still perfunctorily observed, despite an order as recent as 1849).   A master shipwright and an engineer should be attached to the Controller's department and visit each yard quarterly (because the Controller and his assistants were too tied down to do so themselves); thus management practices would be assimilated to the best practice, a uniform interpretation of orders and regulations would be established, the Admiralty would receive valuable feedback, and a better basis for making yard appointments would be established.[18]   It was essential that local copies of the dockyard instructions should be kept current and that each subordinate officer should have for ready reference a copy of those instructions pertaining specifically to his own duties.   The general instructions badly needed to be revised, a task which was taken in hand and completed in 1863-64; check measurement was discontinued in 1863; and the morning meeting was more carefully observed. The Committee's other recommendations were ignored at the time, yet most were eventually adopted.[19]

The system of promotion by merit adopted in 1847 had manifestly failed; the Dockyard Committee, Somerset, Robinson and the superintendents all agreed on that. To make it work, the examinations were henceforth held annually in December; all were prepared, and those for the higher supervisory positions were marked, in the Controller's office; and the superintendents, instead of writing personal reports, gave the candidates points for "personal qualifications" up to one-third of the total number possible. However, it was only by handing administration of the examinations over to the Civil Service Commissioners in 1867 (as had been done with the apprentice examination already in 1860) that the integrity of the merit system was ensured. Nevertheless, complaints about favoritism (arising from the personal marks) were still numerous even in the 1890s.[20]

Should the superintendent have the power to dismiss workmen summarily in cases of serious misconduct? The question had been asked before, but the Dockyard Committee and the Royal Commission asked it again and thereby started a debate. Robinson and the superintendents themselves argued that such power was absolutely necessary for discipline and said, moreover, that they possessed it already and had in fact exercised it. The contrary opinion of Charles Paget, the Admiralty

Secretary, prevailed, however, and in 1864 the superintendents' power of dismissal was limited to hired men. The superintendent might suspend and recommend the dismissal of an established man, but only the Board of Admiralty could impose the penalty itself, a process sufficiently circuitous, cumbersome and uncertain in its outcome to discourage using it very often. Moreover, the men thought of pensions as being deferred wages to which they were entitled, and the superintendents, yielding to pressure from below, had always been, and continued to be, reluctant to invoke the ultimate penalty.[21]

The permanent Establishment, with its entitlement to superannuation, came under fire but survived, although as a much smaller proportion of the dockyardmen. The Dockyard Committee naturally thought it ought to be abolished; after all, private employers made no such provision, nor did the Admiralty itself in the case of the factories. Robinson regretted that the Establishment had ever been instituted yet was under no illusion that it could be abolished now; the men valued it too highly. The argument for the Establishment was that without it a loyal workforce could not be obtained, since the Admiralty paid so little in wages compared to private builders. It was seen also as a defense against unionism, a subject much on the Admiralty's mind in this period. The Treasury grudgingly accepted that a permanent Establishment was needed but not that it needed to be so big, a position with which the Admiralty itself concurred. The Establishment in 1855 had been increased from 9621 to 10,850 and in 1859 accounted for 89 per cent of the workforce. Soon, however, it was being run down (by attrition and temporary closure), to 59 per cent after 1864, at which time it was again fixed at 9621 (out of a total of 18,000 men). Henceforth, dockyard apprentices were placed on the hired list upon coming out of their time, and permanent status became a reward to be conferred upon deserving men and thus an instrument of subordination. The established list by 1890 comprised only 26 per cent of the dockyardmen.[22]

Paradoxically, superannuation, instead of ridding the yards of elderly men (its original, and still ostensible, purpose), had actually lumbered them with such. There was no mandatory retirement age until it was fixed in 1858 at the absurdly high age of seventy, and many elderly men were consequently kept on just to increase their pensions, either out of compassion or because of pressure from younger men mindful of their own declining years yet to come. The age limit was lowered in 1864 to sixty, although hired men might serve until the age of sixty-five.[23]

There was in this period a strong current of industrial unrest and at the same time deep distrust of the workforce. Trouble was avoided, and productivity increased, by reviving piecework when the Crimean War broke out in 1854. The shipwrights were for the time being earning 6s. per day (half again as much as the time rate); still, up to 9s. could be made in private yards and many of the best workers in the eastern yards consequently defected. The dockyards reverted to day work with occasional check measurement as soon as the war was over; this, however, generated so much discontent that the time rates had to be increased (to 4s. 6d. for the shipwrights). Occasional check measurement, however, was soon shown to be "worse than useless" in prodding the men to do an honest day's work, and continuous check measurement was revived late in 1857 (after twenty-four years). The naval program of 1858 followed, and piecework along with it, but both the Dockyard Committee and the Royal Commission having raised serious objections, piecework was again discontinued in 1861, whereupon continuous check measurement was again revived, only to be abandoned four years later when it too was shown to be useless. Meanwhile, industrial discontent was swelling; petitions rained down on the Admiralty and in 1865 became a deluge. These were boom times for the shipbuilding industry and the eastern yards eventually found it impossible to hire enough shipwrights.[24] The iron shipbuilders who had been hired at Chatham struck and were seen off when they refused to go back to work. The hired shipwrights at Woolwich were the only others to down tools but quickly changed their minds when they were threatened with immediate dismissal. Overtime was revived and a premium ("exertion money") of 6d. per day was paid, but only on certain ships, thus creating yet another grievance. The Admiralty, however, did not yield on wages; it was willing to settle for a lower intensity of labor.[25]

The hired men were particularly aggrieved. They were not eligible for a pension and thought, therefore, that they deserved to be paid higher wages than the established men. This in fact is what was eventually done, although not until the next general pay rise, in 1873. In 1864, however, there was an attempt to defuse the issue by agreeing to pay a small gratuity to redundant hired men if they had served continuously for at least twenty years and to extend entitlement to a pension to the widows of hired men killed on the job (hitherto restricted to the widows of established men). What the hired men wanted most of all, however, was to be placed on the established list. In order to improve their chances, the minimum period of service for superannuation was lowered from twenty to ten years, thus making it possible for a man to be established up to the age of fifty and receive, at the age of sixty, the minimum pension (time served on the hired list was not counted). At the same time, the hired men were assured that the new policy of entering all workers initially on the hired list would improve their chances of acquiring permanent status, although the men themselves doubted that it would.[26]

The discontent that had agitated the dockyards evaporated after 1865. The shipbuilding industry was savaged by a long, deep slump and collapsed on the Thames[27] (until now the principal shipbuilding district), while at the same time the Conservative government that took office in 1866 sharply curtailed new construction, the Establishment was temporarily closed, hired men were let go, and the workforce shrank from 18,297 in 1865 to 14,142 in 1868 and (under the no less economy-minded Liberal administration that followed) to 11,276 in 1870.[28]

The arguments both for and against piecework were the same that had been made by the Commission of Naval Revision and the Commission of Naval Enquiry. The system was "open to such objections and such abuses," said the Dockyard Committee, that it could not "with propriety be continued in public service"; there was "no reason" why with good management day work should not produce the same favorable results as in the factories.[29] Robinson, Bromley and the yard accountants fully shared these views, while Walker and the master shipwrights looked upon day work as a system that checked "the spirit of exertion amongst workmen" and tended "to idleness and indifference on their part". It was "only in time of great pressure" that piecework was ever used, Walker said, "to induce men to enter the yards for the sake of higher wages, and to remain in them", and to raise output per man to the requisite level; if piecework were ever to be abolished, the "blow" would "paralyse" the yards. Yet there was no pretence in 1854 that piecework was being revived for any reason other than to increase wages, and the whole business in any case was quite mismanaged. The existing number of measurers (a total of 22 for occasional check measurement) was not increased, nor was a full (although still inadequate) complement appointed until 1856; meanwhile, the men's earnings were reckoned from the inspectors' day books, based in no small part on self-interested information provided by the leading men, while the measurers' own data also were under a cloud, since until 1858 there were no standard rules for measuring work. The officers, for the first time following their instructions, had some work re-measured in 1859 and on one occasion discovered that eight of twenty-six articles had been measured incorrectly. Moreover, piecework was in no small degree self-defeating. A very considerable number of rates were, and continued to be, too high, despite years of experience, a new rate-book (drawn up in 1852-53 and revised in 1854-55), and continuous review of earnings after 1856. The men, knowing that the rate would be reduced accordingly if earnings on any article exceeded a certain amount, were careful not to perform work in excess of that amount. This was particularly noticeable in the case of newly hired men; at first they would perform more work than the others; soon, however, their work habits assimilated to those of the rest. Earnings were expected to average one-third more than the day-rate, that being the quantity of additional work the men were assumed to perform, even though no evidence that they actually did so was ever produced. In fact, average earnings between 1854 and 1856 exceeded the day-rate by 48 per cent, and many experienced men working no harder than they would have done anyway still earned one-third more. The differential narrowed to 39 per cent in 1857, but that was only because the day-rate had been increased.[30]

Most compromising of all, however, was evidence (the first ever) that measuring piecework was "almost...guess work," just as Bromley and Barrow and the Commission of Naval Enquiry had contended long ago. The Dockyard Committee brought to light the curious (although not atypical) case of eight 1462-ton sister ships that had recently been built, two each at four dockyards. The labor for those ships that were built at the same yard cost (with one exception) roughly the same. The differences in the cost of labor between those built at different yards, however, were quite striking — £6321 and £6569 at Chatham, £6450 and £6712 at Devonport, £8621 and £9311 at Sheerness, and £10,065 and £11,119 at Woolwich, the differential between the least and the most expensive vessel being 76 per cent. The Committee, naturally being mystified, obtained detailed records for the *Cadmus* and the *Pearl*, launched respectively at Chatham in 1856 and Woolwich in 1855, and found as follows: 782 articles were charged to the *Cadmus* and 956 to the *Pearl*, 352 were charged to the *Pearl* but not to the *Cadmus* and 178 to the *Cadmus* but not to the *Pearl*, and only 32 of the 604 articles charged to both were identical in value. The dockyard service tried to extricate itself from this embarrassment and variously attributed the discrepancies to, among other things, inadvertent errors either in recording or (especially) attributing certain articles and to the different building specifications of the two master shipwrights. Robinson, however, thought that none of the explanations was "satisfactory in all its parts."[31]

Bromley staggered the Royal Commission when he revealed that the piecework scheme consisted of 51,400 rates for shipwrights' work alone and 94,762 altogether. He did not think that it was possible for anyone to grasp the meaning of his evidence unless they actually saw the scheme itself and accordingly handed in a page for the Commissioners' enlightenment.

> This is the scheme of prices for shipwrights' work... — it only requires to be seen by practical men of business to be condemned, and to be shown to be absurd.... The consequence of this is, that we require an enormous staff of Measurers [at an annual cost of well over £15,000]; our Accountants are over-worked and their Clerks are broken down having to check these things, and the men are not always satisfied; if they complain of their earnings being too low, the next week it is made up to them, and a little more is given to them, and who can check whether it is right or wrong?[32]

Bromley's evidence made a deep impression, but so too did Walker's, and the Royal Commission concluded that even though piecework was "open to great abuse", it might nevertheless be "resorted to in cases of great emergency" so long as it was confined to building (because the work was at least open and regular).[33]

It must have seemed after 1861 that continuous check measurement had become a fixture. Then Robinson and H.C.E. Childers, the Civil Lord (1864-65), demonstrated not only that it was useless as a practical check (the measurers almost invariably credited the men with a quantity of work the value of which exceeded day pay), but that it impaired the yard officials' sense of responsibility and, therefore, was mischievous as well. Robinson wrote:

> I found that according as the Principal Officers were zealous, intelligent, and active, or the reverse, so their opinions varied as to the use of check measurement, the active officers considering it a farce, the slothful advocating it, on account of the shield it threw over their indifferent supervision.

He was indeed able to demonstrate that dear work followed certain master shipwrights when they were transferred from one yard to another.

There was, he concluded, no substitute for "the zeal, intelligence, and practical knowledge of Inspectors, Foremen, and ... Superior Officers".[34]

Measures were adopted in 1864-65, now that check measurement had been discontinued, to provide the yard officials with relief from clerical and other duties that interfered with active supervision. Each foreman of the yard was given a writer to make out notes for the issue and return of stores and to write up the labor expense sheets. The foreman and his writer were provided with a portable office (a "small glazed" enclosure) which afforded a general view of the area where it was set up as well as the convenience of shelter and proximity to the work. The writers also went round the yard at frequent intervals and collected the inspectors' requisitions for stores, thereby relieving the inspectors of the need to present these to the foremen in person. No longer was the leading man personally required to draw stores and thus absent himself from his gang; this was now done by specially designated men, mostly unskilled laborers but shipwrights also in some cases. The leading man was incidentally given additional relief by the establishment of present-use storeholders (really the store cabins abolished in 1822 over Martin's objections)[35] so as to ensure an uninterrupted supply of materials and continuous production.[36]

It was imperative after 1861 to establish control over construction costs. The practice of estimating the cost of building was finally adopted, although at first with little more success than in the case of repairs, which generally exceeded the estimate by a wide margin. The dockyard expense accounts were quite inaccurate, and the master shipwrights only considered themselves responsible for the work itself and not for the cost. It was Walker's opinion that estimates would be useless anyway; "You may give an estimate of the probable cost of a ship, but it is very doubtful whether it would be kept to." Indeed, very few firms as yet estimated costs, and those few found it quite difficult to do, *The Engineer* observing in 1869 that an "estimate right within 20 per cent of the actual cost" was thought to be very good and to reflect "much credit on the engineer and all concerned."[37] J.P. Peake, the master shipwright at Devonport, claimed that he and his assistants, using rule-of-thumb, knew pretty well when they built a ship how much it would cost, since they were "building over and over again the same thing." Still, there had been vast disagreement among the master shipwrights in estimating the cost of the *Vernon* thirty years before. The Navy

Estimates provided for a certain proportion (reckoned in eighths) of each ship under construction to be completed by a given number of men in the course of the financial year. As George Turner, the master shipwright at Woolwich, explained:

> Supposing we have 100 men ordered, and that the ship is to advance two-eighths during one year, that carried on through the whole eight eighths would be the proportion, and then the expense per day of each man would give the probable cost of labour.

The large but uncertain amount of labor required after launching to fit the ship for sea was not calculated, however. There existed no reliable information from which it was possible for the Controller to follow, and thus control, current expenditure, while the Storekeeper General was an independent official and in any case not directly concerned with the management of materiel. The master shipwright signed the annual expense accounts showing the cost of labor and materials (drawn up since 1856 by the yard accountant) merely as a matter of form and accepted no responsibility for their accuracy; that, in his view, was entirely the accountant's responsibility; indeed, Richard Abethell, the senior master shipwright, told the Royal Commission that the question of costs was not in his department, and he did not hold himself responsible. The yard accountant and storekeeper, for their part, were only responsible for correctly recording, valuing and totalling the labor and materials charged to each ship. So inaccurate were these accounts, however, that the Royal Commission concluded that it was impossible to say anything about how much any ship actually cost. An inquiry conducted at Woolwich disclosed that in one eight-month period no less than 7906 errors, varying in amount from 1d. to £400, had been made in various accounts, including 6566 in rating, valuing and totalling timber and stores.[38]

The Royal Commission was led by the evidence to conclude that although dockyard work was executed to the highest standard, it was nevertheless exceedingly expensive and that vast improvements in management were imperative if costs were to be brought down. The best way seemed to be through the medium of the yard accounts, a possibility tentatively explored by Bromley more than a decade earlier and intriguingly developed more fully now, with his concurrence, by the yard accountants at a time when the practice of using accounts directly to assist management was barely a glimmer.[39]

The yard and factory expense accounts were very numerous, far more numerous, indeed, than any private firm would have thought necessary.[40] Their potential value as an aid to management had never been fully understood, however. They were, moreover, inaccurate. This was partly because the clerks who performed the "tedious and difficult" work of compiling them knew that they were never systematically examined (otherwise, it was argued, the inaccuracies would have been detected long ago) or put to any "real use", and partly because they were exceedingly complex and often ill-constructed. Henry C.G. Bedford, the storekeeper and accountant at Sheerness,[41] declared that "a definite system for utilizing the...accounts of expenditure" was absolutely necessary if "the apathy which...truly...exists in the dockyards" was to be overcome. The manner of keeping the expense accounts had to be improved; unity of purpose and control needed to be established. Wage accounts were controlled by the yard accountant and the Accountant General, store and manufacturing accounts by the storekeeper and Storekeeper General; the master shipwright was only nominally concerned with accounts of any kind, and, except for the annual expense accounts and factory balance sheets, the Controller was quite left out. This situation, said Bedford, was an insurmountable barrier to comparing, checking and utilizing the expense accounts and, indeed, to any "systematic plan for improving and assimilating them". Now when a particular question arose, a report was requested, or someone was sent down from London to investigate, but the expense accounts had never been overhauled, nor did any machinery exist for doing so.[42]

The Royal Commission, like Bedford, thought that if the expense accounts were revised and accurately showed the exact cost of every ship, this would lead to lower costs, not least by stimulating competition amongst the dockyards. The revision, a task of simply monumental proportions, was undertaken at first by Bromley, but, his health having given way, he retired in 1863 and the work was carried to completion by successive Civil Lords,[43] James Stansfeld (1863-64) and H.C.E. Childers (1864-65).[44] The revised expense accounts were driven by two imperatives: to provide the Controller and the master shipwrights with information from which they could follow, and thereby control, costs and to account to Parliament in detail for the money voted and spent for ships and other works.[45] The expense accounts were at the same time useful in estimating the cost of building particular ships, although quite a few years passed before sufficient expertise was acquired; meanwhile, the transitional stage through which shipbuilding was passing made estimating doubly difficult.

Although the yard expense accounts of ships and services were sent to the Controller annually, no general account was presented to Parliament until 1860, when the Admiralty, responding to demands for information, began the practice of laying before the House of Commons annually an account showing both direct and indirect expenditure on naval construction for the preceding financial year. Now under Bromley's direction, the classification of labor and store notes was standardized (to facilitate comparison of production costs at the various yards), the form of the expense accounts was revised, and the factory and manufacturing accounts were transferred from the chief engineer to the yard accountant. Local audit clerks were appointed to audit the expense accounts; however, there was no audit of the data on which these accounts were based (e.g. the measurers' records or the inspectors' diaries of work) and thus no real audit of the expense accounts themselves. The expense accounts were put under the Accountant General (as Bromley had proposed in 1848) and an inspector and auditor of dockyard accounts was appointed. (Store accounts, however, remained under the Storekeeper General.) As a result, Bromley in 1863 was for the first time able to charge indirect expense pro rata to each ship based on the value of labor and materials charged to it direct. The following year Stansfeld instituted a stock valuation account; this, however, being based on defective information, was soon found to be of no practical use; nonetheless, it continued to be compiled until 1879. Meanwhile, the parliamentary account underwent a remarkable evolution, from 29 to 270 pages just between 1859 and 1865.[46]

Admirable and impressive though Bromley's innovations were, the consolidation of expense accounts under the Accountant General had a decidedly negative effect on the Controller's ability to control cost. The immemorial (and now revised) annual Form No. 89, Expense of Ships, and its monthly counterpart, were the only means the Controller had of following expenditure; these, however, the dockyards now sent not to him but to the Accountant General. The Controller no longer received the monthly returns at all and the annual returns, which he formerly received from the yards three to six months after the end of the year, the Accountant General now forwarded about a year after; thus up to two years might elapse since the works had been completed, and the amounts authorized and actually spent could be compared, by which time it was too late to do anything.[47]

Robinson tried desperately to control spending for repairs and in 1864 secured the appointment of an inspector and valuer of dockyard work to examine ships designated for repair, examine the local estimates item by item, put his own value on the work, and cooperate closely with the inspector and auditor of dockyard accounts. It was like scaling Mount Everest, though, for the professional officers had never given any thought, in Robinson's words, to "the probable expense of any given work, and...[had] been practically dissociated from any knowledge of the actual expenditure they incurred." With such nonchalance were estimates prepared that they often had to be returned and explanations had to be obtained. Robinson wrote in exasperation that:

> it required circular upon circular from the Controller, reprimand upon reprimand from their Lordships, accompanied by threats of dismissal, to bring the Officers at all the Yards to a sense of the paramount duty of watching that expenditure should not exceed the estimate without obvious and sufficient cause.

The yard officers and the Controller beginning in 1865 were supposedly provided with full information about both past and current expenditure. The superintendent was notified before the beginning of the financial year of the money voted for dockyard and factory wages for his yard and of the proportion that might be spent weekly. The Controller received a new weekly return ("practically the basis of the Dockyard Returns No. 89") which was intended to enable him to follow the expenditure of labor continuously in relation to both the parliamentary grants and estimates for repairs. This return was structured, however, for parliamentary rather than the Controller's purposes, nor was it forwarded in a timely manner. Moreover, the master shipwrights usually did not send estimates of repairs until after the work had commenced, and sometimes not until after it had actually been completed. It was at this same time that the practice was adopted of charging indirect expenses to each ship in proportion to direct expenditure, "in accordance with the principle of account observed in the commercial world." Thus did the Admiralty hope to be able to compare the cost of building in the dockyards and private yards. Dockyard work, however, was so much more varied than that performed in private yards that the allocation of overhead charges was bound to be problematical. The practice was carried too far, as the Admiralty soon discovered to its dismay, and was subsequently refined.[48] Nevertheless, an Admiralty committee reported in 1884 that the whole question of incidental expenses was "so obscure as to render unreliable any comparison between the cost of shipbuilding in public and private yards. "[49]

It was Childers's opinion that:

the present system under which the preparation and supervision of accounts (not being accounts of quantities) is divided between the storekeeper, the accountant, and the audit clerk of the yard...[causes] much unnecessary writing, copying, and double computation; [renders]...unity of responsibility impossible...; and...[is] calculated to produce inaccuracies.

Thus the duties of the audit clerk in 1865 were transferred to the yard accountant; the result, however, was the mere semblance of an audit. Childers thought it unsound practice also for both the preparation of the wage accounts and the payment of wages to be under the yard accountant, and a yard cashier was appointed for the latter. The master measurer's office was abolished; what remained of its statistical work was transferred to the accountant, and the recording of work (the accounts were expanded) was transferred to the inspectors (their number was accordingly increased). The inspector and valuer of dockyard work was reinforced with two assistants who randomly examined work and ascertained whether the workmen's wages had been certified properly. Finally, the store receiver's office was abolished (he was an inefficient check on the storekeeper) and his duties were transferred to an inspector of stores under the yard accountant.[50]

Childers's achievements as Civil Lord led to his translation to the Treasury, where as Financial Secretary (1865-66) he was largely responsible for the Exchequer and Audit Department Act of 1866, "tying together the procedures of Estimate, Appropriation, Expenditure and Audit in one coherent system,"[51] rather as had been done already at the Admiralty.

Questions in the House of Commons about naval expenditure were becoming more frequent and difficult to satisfy.[52] Highly embarrassing revelations were made in evidence before the Select Committee on Admiralty Monies and Accounts in 1867-68 and were an important factor in the Conservative government's decision to build no more ships in the dockyards for the present (reversed by the Liberals when they returned to office). One case (quite like that of the eight ships a decade earlier) concerned the extraordinary differentials in the cost of building six sloops of the *Amazon* class (1864-66) the dearest of which cost 26 per cent more for labor and materials (46.5 per cent more for labor alone) than the cheapest. This was partly explained (as in the earlier case too) by errors in recording or attributing certain items or charges. It is clear in the case of the most expensive ship, however, that the master shipwright permitted the workmen to idle their time away. No less embarrassing was evidence that it cost considerably more — £4 per ton (8.5 per cent) — to build and convert ironclads in the dockyards than in private yards. Robinson argued, however, that the cost of dockyard-built ships was exaggerated by excessive indirect charges and was able to bring the differential cost down to 5s. per ton. Moreover, he pointed out that contract prices were misleading, since private builders often billed the Admiralty for additional expenses long after the ships in question had been delivered. It was Robinson, in fact, who presently got the Admiralty to adopt fixed-cost contracts. Still, he knew that dockyard-built ships were dear, even though he dared not say so publicly, and regarded bringing costs down as the greatest challenge facing the Admiralty.[53]

The proceedings of the Select Committee (Charles Seely was chairman, and Childers was a member) acquired a political complexion and became an indictment of the Duke of Somerset's administration of the Admiralty. It was Seely's conclusion that the state of the expense accounts made it impossible to ascertain the exact cost of any ship built or repaired in the dockyards (the same conclusion reached, as has been seen, by another committee sixteen years later). So upset was Robinson by the "Very grave accusations [which] are brought forward day by day" and "received with avidity," and so concerned that unless the Admiralty's case was put "in its true light..., very great discredit will attach not only to former Boards but to the present Board of Admiralty," that he procured the appointment of a barrister to observe the select Committee's proceedings and advise the Admiralty. The Committee, however, rejected Seely's pejorative report and adopted Childers's milder report instead.[54]

The problems that Seely's committee had brought to notoriety were seen as being essentially problems of accounting. Thus the Navy Estimates were retabulated, and a new scheme of unified accounts was drawn up accordingly.[55] The various accounts underlying the Estimates were now, however, "made out more from a parliamentary and outside point of view," rather than, as previously, from a strictly departmental viewpoint. Each yard was treated as a separate establishment and accounted for the money "disbursed on its behalf in the final Expense and Manufacturing Accounts". Thus, it was claimed, the relative economy of each yard could be compared, and the Admiralty would have valuable detailed information (provided by three new monthly returns showing, respectively, the cost of hulls, etc., engines and boilers, and indirect charges) hitherto lacking. Little use was made of these voluminous returns, however, for although they were only good for following progressive expenditure, they were not received until at least two weeks after the end of the month, and the professional officers relied instead on rough notes (which might or might not be accurate) kept by the foremen of the yard's and other writers. Moreover, estimates of repairs continued to be forwarded in many cases after the work had either commenced or been completed.[56]

Except in the dockyards, it was not the shipwrights but rather specialized metal-workers (originally engaged in boilermaking), of whom the platers and their helpers were the most important,[57] who dominated shipbuilding in the age of iron and steel. That the platers should have supplanted the shipwrights was only partly owing, however, to their skill in cutting and bending metal plates to shape. The shipwrights might have learned metal shipbuilding also — that would not have been too difficult, even though the material and the tools were quite different from what they were accustomed to — but so much did they dislike working in metal that they chose to remain workers in wood alone. Unskilled men were hired also and taught to be riveters, holders-up,[58] etc., whereupon they either were organized by the United Society of Boilermakers and Iron Shipbuilders (to which the platers belonged) or formed their own unions (eventually absorbed by the Boilermakers) and won recognition as mechanics.[59]

It was platers that the Admiralty naturally hired when iron shipbuilding commenced in the dockyards. But, as we have seen, the platers struck, as part of a national movement by their now powerful and militant union to raise wages, and were dismissed in 1863. The dockyard shipwrights, not wanting to suffer the fate of shipwrights elsewhere, seized the opportunity and learnt metal shipbuilding, even as the shipwrights nationally were trying (alas, too late) "to establish their right to work on metal construction." Thus did the dockyard shipwrights become invaluable "all-round men" — "the backbone of the service" — working not only in wood but "in iron, steel, and every conceivable thing incidental to the building of the hull and fittings of a man-of-war...undertaking, for example, plating, armor-plating, and torpedo fittings," work which elsewhere was performed by mechanics belonging to various specialized trades. A new category of workers, the skilled laborers, was created by teaching certain unskilled laborers riveting, holding-up, iron-caulking, and so on, and paying them a slightly higher wage. This structure of the workforce, unique to the dockyards, which developed beginning in the 1860s was highly advantageous to the Admiralty. It kept wages low (e.g. the boilermakers had been paid 7s.-8s. per day, whereas the shipwrights got a mere 4s. 6d.); the shipwrights could be shifted readily and economically among various kinds of work according to need, unlike specialized mechanics (e.g. platers, whose work was specific to shipbuilding, could not be employed economically on repairs); and the curse of demarcation disputes was largely avoided. By the same token, however, bitter seeds of discontent were sown. The shipwrights were misled to expect that their wages would be increased in consideration of the new skills they had undertaken to acquire, while the skilled laborers eventually wanted to be classified as mechanics, and to be paid more, as men doing similar work in private yards were; this latter, however, was a grievance the Admiralty never redressed.[60]

The structure of the workforce, of course, was not to the advantage of the Boilermakers' Society, since it practically shut them out of the dockyards. They might have exploited the men's grievances to their advantage but until 1902 chose to boycott the yards instead.[61]

It was not at all easy to find at the Admiralty a naval architect who had a good grasp of theory and "was ready to adopt the new improvements or iron-shipbuilding in all its ramifications, and of armour-plated ships." The theoretical grasp of the dockyard constructors was even weaker. John Scott Russell, the designer and builder of the *Great Eastern* (1858),[62] had to be employed, in an unprecedented collaboration, to help Isaac Watts, the first assistant surveyor, design the *Warrior*. Watts was subsequently reluctant, however, to design either composite or iron ships, or even to propose how design might be improved. Somerset and Robinson lost patience and Watts was sacked in 1863. He was succeeded by Edward Reed (with the title of Chief Constructor), one of his assistants and a graduate of the second School of Naval Architecture. Reed was strongly recommended by Robinson but encountered considerable opposition anyway; he was only thirty-three years old, and three wooden sloops were the only ships he had designed. Nevertheless, he was unquestionably the right man for the job, a firm and persuasive advocate of iron-hulled over composite ships and the author a few years later of *Shipbuilding in Iron and Steel* (1869), the standard work until William White's *Manual of Naval Architecture* (1882).[63]

The Admiralty constructive staff, consisting of two contructors besides Reed (for dockyard correspondence and inventions), was enlarged by three more to assist Reed with the drawings and questions of detail that arose during construction. Draftsmen until then had provided the chief constructor's only assistance in preparing detailed drawings. That had not been the best practice but was tolerable so long as ships varied but little in their essential parts. Now, however, that naval construction was in a transitional state and evolving very rapidly, all kinds of research and calculations, hitherto unnecessary, were imperative.[64]

The establishment of the Royal School of Naval Architecture and Marine Engineering in South Kensington in 1864 was at least as much the consequence of the deficiencies discovered in the constructive staff as of the campaign by the Institute of Naval Architects (founded in 1860 by Scott Russell, Reed, Robert Barnaby[65] and others). The school (it merged with the Royal Naval College, Greenwich, in 1873) produced the greatest naval architects of the late nineteenth and twentieth century and provided the Admiralty with a corps of highly educated constructors in marked contrast to commercial builders most of whom, even as late as 1914, employed few, if any, similarly educated men. Admission to the school, by examination, was chiefly from the dockyard apprentice schools the standards of which had recently been raised. A few private students were admitted also, 60 (mostly foreigners) of the 166 who completed the course down to 1914, only six of whom were employed by private builders, however.   Chairs of naval architecture were subsequently founded at several universities (Glasgow, 1883; Armstrong College, Newcastle-upon-Tyne, 1890; Liverpool, 1908), but private firms were mostly indifferent, or even hostile, to educated men.[66]

A further stage with either new or deeper insights had been reached in the slow modernization of organization and management. The impetus for change was mostly external, however, political (the increasingly prominent parliamentary factor) and financial (the Treasury was as close with a penny as with a pound), rather than internal. Walker took the initiative in insisting upon a delegation of authority to the Surveyor, but otherwise there was no inclination, either at the Admiralty or the dockyards, to alter the existing system. Change there was anyway, not radical, to be sure, but still important and, of course, the work of many, especially Robinson, Bromley and Childers. Robinson was the leading player, however, and, significantly, an outsider, as was Childers also.

The enormous expense associated with the revolutionary innovation of steam and armor and with a vast naval program naturally focused parliamentary attention on the dockyards. The House of Commons was insisting that the public should receive value for its money, and more than a little suspicious that it was not by a long chalk and that private yards could build warships more cheaply than the royal yards. The Admiralty responded with the Committee on Dockyard Economy, but its own committee opened a Pandora's box, pronouncing such strictures and causing such a storm as to make the Royal Commission all but inevitable. Yet the Royal Commission reached the same conclusion: the dockyards on the whole were not well managed. It was the system that was blamed, although individuals were not spared. Both the Committee and the Royal Commission adopted the same view as Bentham, that the dockyards should be treated the same as other large manufacturing establishments and be judged by the same standards of performance. The Committee focused on local management, while the Royal Commission took a larger view and concluded that the Admiralty itself needed to be reorganized. There was no unity of direction and control, and communication among the various departments and coordination of their separate, but related, activities were problematical at best. Nonetheless, the system of management, with one exception (delegation of authority to the Controller), remained recognizably the same as in the eighteenth century.

The harshest judgement was that although the quality of work was excellent, the cost was excessive. Better information was correctly perceived to be a key to reducing the cost and increasing the efficiency of production, yet there was in practice remarkably little to show for all the intelligence and labor that were brought to bear on the subject. The multifarious dockyard and manufacturing accounts (and factory accounts also) were revised and made more intelligible and reliable, originally with a view to mobilizing them as an aid to management as well as providing the House of Commons with full information. In the end, however, the accounts were driven by the political imperative of satisfying the House of Commons that the money voted by Parliament had been spent exactly as Parliament had specified, rather than by the needs of management, and thus were of little use to the Controller. Nor did the revised expense accounts stimulate the salutary competition amongst the yards that was anticipated. Calculating and adding indirect expenses to the cost of production was thought to be a great achievement, one that would enable comparisons to be made with private yards, but the calculations were badly flawed, and the practice backfired. Still, accounting practice and procedure were much improved. The institution of an audit, since it was merely internal, conducted by the accountant's office, and did not include the underlying accounts, was undoubtedly a step in the right direction, even if it was not very effective.

Robinson relied on the inspector and valuer of the dockyard work (a position created through his efforts) and on admonition and reprimand to control the cost of production, but most of the master shipwrights had never given a thought to the cost of anything, and the habit of a lifetime, institutionalized by a system that denied local managers any latitude to take independent decisions, was not easily changed.

Bromley exposed the absurdity of piecework and thereby put a hoary controversy to rest for a generation. The main purpose of piecework, however, had never been to increase output per man above some notional norm but to increase wages temporarily. Wages were a serious but inadequately redressed grievance that clearly affected productivity. Indeed, the issue of wages (or, rather, of wage differentials), now that the structure of the workforce was undergoing radical transformation, was a time-bomb waiting to go off.

By no means least important, the decisive role of the human factor in management was grasped. The effectiveness of supposedly self-acting and essentially coercive instruments like check measurement in controlling costs and increasing output per man was exploded. It was seen that there was no substitute for "the zeal, intelligence, and practical knowledge" of managers and supervisors. An impressive range of measures — tightening the standard of admission and enhancing the curriculum of the apprentice schools, the new School of Naval Architecture, refinement of the merit system, and relieving supervisors of certain clerical duties — was instituted in the expectation that the quality and effectiveness of both management and supervision would thereby be enhanced. It was only in the long run, however, that the results of such measures could be known.

Chapter 7

Revolution and Counterrevolution, 1868-1885

Childers returned to the Admiralty as First Lord in Gladstone's first administration in December 1868 determined to confront the challenge of bringing the cost of building warships in the government yards in line with that in private yards and at once embarked upon an audacious and controversial program of structural reorganization. In the space of little more than a year the Controller's department was reorganized, the Controller himself was made a member of the Board of Admiralty (as Controller and Third Sea Lord), and the Storekeeper General was subordinated to him, local dockyard management was restructured, and business methods and individual responsibility were introduced. Meanwhile, Childers continued the late Conservative administration's policy of retrenchment but reversed another of their policies and started building ironclads again; the next sixteen years, nonetheless, were for the navy a period of relative neglect. The biggest savings were made by closing Woolwich and Deptford yards,[1] an overdue measure (and already in the works following the modernization of Chatham) which largely made possible a further reduction of the workforce, from 14,142 in 1868 to 12,850 in 1872.[2] The yard inspectors were abolished, on the grounds that they constituted an unnecessary supervisory layer,[3] and lower-paid writers were substituted for temporary (in some instances established) clerks.[4]

Childers, however, suffered a nervous breakdown and resigned in March 1871, his radical innovations only partly executed and in trouble, and the Admiralty itself in turmoil. Two years of driving reorganization and autocratic rule were now followed by five years of reappraisal in the course of which Childers's innovations were either modified or undone. It is hardly surprising that, except for the creation of the Royal Corps of Naval Constructors in 1883, the next nine years were not very remarkable. Indeed, Childers's successors themselves were mostly unremarkable. G.J. Goschen (1871-74) was the exception, but having been promoted from the Presidency of the Poor Law Board lacked Childers's rich experience, acquired while Civil Lord of the Admiralty and Financial Secretary to the Treasury; moreover, Childers was soon back in the Cabinet, as Chancellor of the Duchy of Lancaster (1872-73), which cannot have made Goschen's job of tidying up any easier. Neither G.W. Hunt (1874-77) nor H.W. Smith (1877-80), in Disraeli's administration, was a notable First Lord, nor was the Earl of Northbrook (1880-85) in Gladstone's second administration.

Childers's breathtaking coup largely embodied, although it did not in all respects faithfully reflect, ideas developed by Sir Spencer Robinson and rejected by the Duke of Somerset as being too radical. Those ideas, incisive and seminally important, Robinson in 1867 laid out in a memorandum originally intended for his own use and later placed in evidence before the House of Lords Select Committee on the Admiralty.[5]

Robinson spoke in the memorandum of a record of failure, of how Graham's reforms, the reports of the Committee of Dockyard Revision, the Committee on Dockyard Economy and the Royal Commission, and the endeavors of successive Surveyors of the Navy had all failed to put the management of the dockyards and of naval programs on a sound footing. Why was this so? he asked. Because, he said, the dockyards were great manufacturing establishments and should be, but were not, managed according to the principles followed by successful manufacturers.

> It is undeniable that in conducting even one large establishment, in which many various trades are carried on, and large bodies of workmen are to be employed,
> method, forethought, and unity of purpose, are above all necessary to direct the work;
> Accurate accounting of the expenditure, and of work done, to judge of its economy;
> Personal and unified responsibility, to the supreme authority in the

conducting agent;

And complete control on his part over all without exception, through whose instrumentality he works, to ensure the due execution of the instructions he receives.

If this be true for one establishment, how much greater is the necessity of such an organization for the proper control and management of seven such establishments?

As to the first principle, Robinson wrote, it was the Controller's function, in part, to direct naval construction in each dockyard, yet his hands were tied; he was completely subordinate first to the Board of Admiralty, second to his superintending lord, and third "to any Lord superintending a department which has any connection with dockyard expenditure. "

> *On all points* the decision of the Superintending Lord over-rules the Controller; orders are sent to the dockyards, and reports and observations are made in reply, which need not go through the Controller's hands, or with which he is not acquainted until action upon them is begun.

> The Storekeeper General also exercises an authority in the dockyard entirely independent of the Controller; in a word, the management of the dockyards is vested in the Board of Admiralty, in the senior Naval Lord, in one of two Junior Naval Lords, in the Controller of the Navy, in the Storekeeper General, and in some respects, though not directly as to the disposition of work, in the Secretary of the Admiralty.

> How can there be unity of purpose, forethought or method in conducting these establishments?

The Controller was virtually powerless, Robinson said in the memorandum. It was impossible for him to prepare a naval program that would not be interrupted and delayed. The First Lord might very well order work for which no provision had been made and, indeed, might even give a direct order without the Controller knowing about it for days. A ship's commander might insist upon the performance of work of doubtful necessity. The interests of the Controller and the First Sea Lord were in fact at variance. For it was the interest of the Controller to build and repair ships as economically as possible, whereas the First Sea Lord, having the efficiency and strength of the fleet as his first concern, had an interest not in controlling expenditure but in increasing it. There was, moreover, no coordination between the Storekeeper General, who supplied the materiel, and the Controller, who planned the work.

Robinson thought the Controller, hitherto a servant of the Board of Admiralty, should himself be a member of the Board.

> His connexion with his colleagues, and with the First Lord, will keep them much better informed about dockyard matters than they can be now; the authority in the dockyards will be single and not manifold as it is now; the views and opinions of his colleagues will have a direct influence upon him...; nothing will be done without his knowledge; nothing will be decided without his being heard; the frightful correspondence that passes between the Controller and the Board will be extinguished; the delays in the transaction of business, the loss of papers, the want of due information on subjects connected with the dockyards, and with the designs and fittings of ships now experienced, the constant misunderstandings which occur in executing by one department approved submissions emanating from another department...would disappear at once.

The biggest problem, in Robinson's view, was the organizational structure of the dockyards themselves. The magnitude of the problem was strikingly revealed, in 1872, in the Report of the Royal Commission on the loss of the *Megaera* (a troopship converted from an ironclad) on her way to Australia the year before. The Royal Commission concluded that the vessel was unseaworthy and severely criticized the officers at the yard where she had been converted. "[They had done] no more than each of them thought it was absolutely necessary to do; following a blind routine in the discharge of their duties, and acting almost as if it were their main object to avoid responsibility."[6] That was a verdict with which Robinson did not disagree.

> The whole object of my life as a public officer has been to bring together the public departments...; my object has been that the master shipwright and the chief engineer, two independent functionaries in the dockyard, should understand that they are to work together. There is memo. upon memo. of mine, and private exhortation after private exhortation, that these officers are not to consider themselves only responsible for the bare technical duty they had to do, but to consider that they were jointly responsible for all that passed...until you get under one head the whole management of a dockyard you will never succeed in being perfectly certain that some one has not omitted something, because it was not in his department according to his judgement.[7]

Each yard, Robinson said in 1867, must have a general manager to coordinate and control the constructive, engineering and store departments but had instead a superintendent who was responsible for nothing that was

> either done or left undone...so long as he duly transmits the memoranda from the Admiralty, the Lords, the Secretary, the Controller, the Storekeeper General, The Accountant General....
>
> A work which ought to have been done for 10,000*l.* costs 16,000*l.*, he is not called on to account for this excess; he has no responsibility on account of it. When a question is asked, he directs the master shipwright to reply.
>
> If...any materials required for the works in his dockyard are not forthcoming, and the service is delayed, and rendered more costly thereby, he is not responsible; he turns to the Storekeeper, and transmits that officer's replies to any question that may be asked.
>
> He is equally without complete control over those through whose instrumentality...[he] professes to work; he cannot promote, he cannot dismiss, he is checked by antiquated instructions.
>
> If there were not waste, if there were not mismanagement, it would be a miracle.[8]

Childers took office on 2 December 1868; presiding at the Board the next day for the first time, he announced a sweeping reorganization of the Admiralty. The organizational structure of the Admiralty had been a subject of debate for a number of years, and although Childers rejected the argument that the Board served no useful purpose and should be replaced with a minister of marine, he nevertheless created a departmental system of administration reflecting mid-nineteenth-century distrust of collective management by boards and preference for unambiguous lines of individual responsibility. The First Lord now became unambiguously "the superior authority," Childers declaring that he had accepted office

> on the clear understanding that the responsibility should be placed distinctly on myself, as the Minister responsible to the Crown and country for carrying out the public wishes; and those who are associated with me at the Board of Admiralty fully recognize this fact, and will, while I am in office, perform their share of the departmental administration in direct subordination to myself.[9]

The business was divided among three departments under the junior lords as follows: the fleet under the First Sea Lord (assisted by the Junior Sea

Lord), materiel under the Controller and Third Sea Lord, and financial administration under the Financial Secretary (assisted by the Civil Lord).[10] Thus began the new First Lord's autocratic, and ill-fated, rule.

Childers and Robinson, working closely together, proposed, through a parallel reorganization of the Controller's department in London and the dockyards, to make the royal yards competitive with private yards.

> The amount spent [in the dockyards] was so immensely in excess of that expenditure in private yards that we naturally inquired as to whether there were sufficient reasons for so great a difference; of course, under any circumstances, the cost of a naval dockyard would be greater than a private yard, and the nature of the work differs so much that an accurate comparison is hardly possible; but the question was whether the expenditure...was justifiable, and whether we could get the same or greater efficiency by a consolidation of offices....[11]

Making the Controller a member of the Board of Admiralty and downgrading the store department to a branch of his department (under the Superintendent of Stores) were highly controversial measures. The object was to unify control, simplify the conduct of business, and provide a single and complete record of everything pertaining to ships and ship construction. (The purchase of stores was at the same time hived off and became a branch of the Financial Secretary's department.) The consequent increase in his work was more than offset, Robinson said,

> [by making] the work of the dockyards so much lighter because I was able to see that the proper provision of stores actually wanted was in the several yards, whereas before I had to correspond with the Storekeeper General and the Storekeeper General had to go to the Third Sea Lord [his superintending lord], and the Third Sea Lord had to turn it back to me.[12]

Coordination was improved by amalgamating the Admiralty constructive and engineering staff under the Chief Constructor assisted by an engineer (the Surveyor of Factories and Workshops and Consulting Engineer). The Chief Constructor was generally responsible for anything related to production, controlled demands for materiel and corresponded directly with the superintendents on matters of professional detail. Control over production and costs was strengthened by appointing two examiners of dockyard work (in place of the Valuer and Inspector of Dockyard Work appointed in 1865).[13]

Robinson's own plan for local management was that the dockyards should be organized like private yards and that each should have a general manager to coordinate and control ship construction, engineering and stores. The master shipwright and chief engineer often worked at cross purposes, sometimes even in a spirit of antagonism, while disputes between the master shipwright and the storekeeper were "perpetual" and production was often held up at great expense because the requisite materials were not on hand (so little effective was the board of estimate).

> The master shipwright had no authority to order the [storekeeper]...to procure and keep [the stores]...in the way he thought necessary for his work, and although he ought to have been in thorough communication with him on these subjects he often was not. The system of coequal authority led to carelessness on the part of the master shipwright; often due forethought was not exercised in making demands; it was easy to say, I have been waiting for materials, and...[for the storekeeper to reply], I was not asked in time or in a proper shape.

At some yards stores were permitted to accumulate beyond what was needed and obsolete stores continued to be ordered; articles differing so slightly from those actually specified as to be interchangeable were allowed to become surplus or obsolete or to deteriorate; storehouses were encumbered with unserviceable or surplus stores; stores were arranged without system or method. Indeed, despite inventory with a nominal value of £13.5 million, an additional £200,000 had to be spent immediately in order to bring forward the requisite number of ships at the time of the Franco-Prussian War in 1870. The problem was unsystematic and uneconomic management; this Robinson attributed to the storekeepers' lack of shipbuilding experience.[14]

In Robinson's scheme, the general manager, assisted by the master shipwright, engineer, and storekeeper, was to organize and coordinate naval construction and engineering, allocate labor and materials, and control expenditure. He was to be a civilian, not a naval officer. The position required a man possessing "much method, foresight, and an immense amount of technical knowledge" such as, in Robinson's experience, the education of a master shipwright generally did not afford; thus he would have appointed a chief engineer. Robinson did not doubt that a naval officer (although not of flag rank) possessing the requisite qualifications might be found, but, he observed, appointment as a dockyard superintendent "was looked upon as a reward for other services".[15] Still, the superintendent's office was not to be suppressed, for not only was "a naval link...necessary between the commander-in-chief at the port and the dockyard," but a dockyard had certain functions other than building and repairing ships which were "especially within the province of a naval officer".[16]   This, of course, is exactly the organizational structure that was eventually adopted, although not until the twentieth century was well advanced.

It was not Robinson's ideas that were carried out but Childers's version of them. The scheme naturally encountered heavy opposition and for this reason was introduced experimentally in 1870 at just two yards — Portsmouth and Chatham — with the intenting of adopting it generally in a year or so, when, it was expected, opposition would have died down. The engineering and storekeeping departments were suppressed and consolidated with the constructive department; thus the master shipwright became general manager (with the additional title of engineer).[17] The assistant managers (for ship construction, engineering and storekeeping) were not appointed; the master shipwright and engineer did it all. Store business was very heavy and technical, however; thus a foreman of the yard was appointed to be the master shipwright and engineer's executive assistant in charge of stores and storehouses, while the rating and valuing of store notes and the preparation of the stock valuation accounts (instituted in 1864) were transferred to the yard accountant, so as to provide an independent check. Nevertheless, there was considerable doubt about the wisdom of annexing the store department, and that arrangement was to be reviewed after six months. Five years passed, however, before it was.[18]

By 1871 the reforms, launched with such high expectations, were unravelling, Robinson was on the way out, and Childers was a broken man. The Board of Admiralty all but ceased to function as such under the departmental system and Childers's autocratic regime. Its meetings were infrequent (only thirty-three in 1870)[19] and perfunctory, usually lasting only a few minutes. It was Childers's custom to take decisions without consulting the Board as a whole; there was no opportunity for general discussion or an exchange of views; the junior lords administered their departments and advised the First Lord in isolation from each other, and Childers sometimes took decisions behind Robinson's back.[20]

There was particularly sharp disagreement over ship design, between Edward Reed, the Chief Constructor, supported by Robinson, and Captain Cowper Coles, who had the ear of Childers and the naval lords.[21] Coles was responsible for HMS *Captain*, a full-rigged turret ship designed under his supervision by Lairds of Birkenhead and the only warship the Admiralty ever built that its own staff had not designed.[22] The *Captain*, however, was a very unstable ship; on the night of 7 September 1870, while on her maiden voyage, she capsized in a squall in the Bay of Biscay and nearly all on board, including Coles and one of Childers's sons, were lost. Childers bitterly blamed Reed and Robinson for the disaster, even though both had expressed deep misgivings about the ship's stability. Reed resigned.[23]

The amalgamation of dockyard offices was running into problems, even though the master shipwrights were predictably enthusiastic, and warnings, which proved to be well founded, were ignored. For example, the superintendent at Chatham warned that the new inventory-taking arrangements were too great a burden for the master shipwright and engineer in view of the many technical problems with which he had to be concerned because of rapid changes in ship design and construction.[24] The Franco-Prussian War, necessitating precautionary mobilization in the middle of still undigested organizational changes, was simply a spot of bad luck.

Childers went to pieces from nervous exhaustion, brought on by overwork, dissension at the Admiralty, and the loss of his son in the *Captain* disaster, making it impossible to approach him on difficult questions which only he could decide.[25] Early in 1871 he retired briefly to Madeira, hoping there to recover his health, but returned to England in a state so unfit that he resigned.

Meanwhile, Robinson was forced out, the reorganization of the dockyards was sinking deeper into trouble, and the Admiralty was demoralized and in disarray, the subject of an inquiry by a select committee of the House of Lords chaired by Childers's and Robinson's old chief, the Duke of Somerset. It was the task of Childers's successors to deal with his untidy legacy.

Robinson was succeeded by a stop-gap, Captain Robert Hall (1871-72), the superintendent at Pembroke, and he in turn by Rear Admiral William Houston Stewart (1872-81). Stewart had acquired a wealth of experience as superintendent at Chatham (1863-68), Devonport (1868-71) and Portsmouth (1871-72) and was, except for Hall, the first Controller to have had dockyard experience, but although he was an efficient administrator, he had little of Robinson's intellectual power.[26]

The House of Lords Select Committee, reporting just as Childers resigned, was sharply critical of his controversial innovations at the Admiralty. The Report itself was just as controversial, however, and was agreed only because the Committee divided equally along party lines. Childers was accused of having "so far disabled the Board that it is no longer fit for consultation or for the review of naval affairs." Goschen did not forsake the departmental system; nevertheless, under him the Board met twice a week and the junior lords met daily to read the more important correspondence and exchange information of general interest.[27]

It was the curious opinion of the Committee (or so it much seem today) that it had been a serious mistake to elevate the Controller to the Board of Admiralty, thereby putting him "in an anomalous position, since he was at once a member of the Board of Admiralty and also serving under the Board of Admiralty." He was removed in March 1872 but continued to be directly and solely responsible to the First Lord and attended the Board, whenever anything pertaining to his department was under consideration, and the daily meeting of the junior lords also. Moreover, in order to emphasize the principle of individual responsibility, he henceforth signed the more important letters emanating from his department, while the heads of the various branches signed the rest, rather than the Admiralty Secretary "by order of their Lordships."[28] The Controller's position, therefore, was not so very different from what it had been under Childers, although it was clearly less satisfactory. The Controller was restored to the Board in 1886, however, and Childers and Robinson were vindicated.[29]

Naval design and construction, marine engineering, and the management of the dockyards were highly specialized and functionally distinct, albeit coordinate, activities and more than one person could well handle in the new age of shipbuilding. This was reflected in the organizational structure of large private firms that Goschen wisely adopted for the Controller's department in 1872. The Chief Constructor became Director of Naval Construction,[30] the position of Engineer-in-Chief was revived, and that of Surveyor of Dockyards was created (a highly promising measure). Coordination was by the Council of Construction (to which Childers had entrusted the Chief Constructor's duties when Reed resigned), representing all three branches and chaired by the Director of Naval Construction. It was an arrangement with which a decade later the Council itself expressed general satisfaction.[31]

The Surveyor of Dockyards, Frederick Barnes (1872-86), was responsible for "the maintenance and repair and cost of building ships and engines as well as for the details of the management of the dockyards," including coordination of the construction and engineering departments. An anomalous vestige of the old, undifferentiated arrangement survived, however. For the Surveyor of Dockyards communicated with the Controller through the Director of Naval Construction, Nathaniel Barnaby (1872-85),[32] and Barnaby's staff handled the administrative work; Barnaby, however, never interfered with Barnes. Assisting the Surveyor were an Admiralty constructor (the former professional assistant to the Chief Constructor), an engineer, and three examiners of dockyard work (an increase of one). The examiners' role in controlling costs was crucial; no repair or refit exceeding £2000 or £3000 was undertaken without one of them having inspected the ship and gone through the estimate item by item with the foreman in charge. Inspections were often ordered even for work involving much smaller amounts (sometimes as little as £100) if the estimate appeared to be too high or there was a question about whether the money should be spent. It was expected (even though it did not work out that way) that the Surveyor himself would frequently visit the yards so as to exercise a wider control and provide assistance and advice to the local officers.[33]

Childers's controversial amalgamation of the master shipwright's and engineer's departments at Portsmouth and Chatham was not reversed until 1875. Barnes, Stewart and the Board of Admiralty all agreed that it was "absolutely necessary" to coordinate the two departments, and that the master shipwright was the proper person to do so, but thought that giving him the additional title and responsibilities of engineer had been a mistake, leading him, in Barnes's words, "to consider himself the highest authority...on all matters [of engineering]", even though in fact he had no competence in most. The compromise solution was to make the engineer "in all respects a Principal Officer of the Dockyard" and not subordinate to the master shipwright; "*in all questions of management*", however, he was "to be subject to the concurrence" of the master shipwright, and the master shipwright was "to have the first charge of all the work and in the shops in regard to *time of completion and to general arrangements.*" It was at this time that the master shipwrights and Admiralty constructors were given the title of constructor and the chief draftsmen at the Admiralty that of assistant constructor.[34]

Dissatisfaction over wages, among workmen and officers alike, ran high in the 1870s; morale sank very low and must have affected the way people did their jobs. Discontent amongst the workmen had been building up since the 1860s. The shipwrights had not gotten the rise that was promised when they agreed to learn iron shipbuilding, and wages in private shipyards meanwhile had risen. The shipwrights and most of the other men got a rise in 1873 and the rest in 1875. The shipwrights' wage rose from 4s. 6d. to 5s. per day, yet they were not given the higher differential wage to which they thought themselves entitled, a grievance which festered for many years.[35]

The constructors had even more reason to be dissatisfied. Their salaries had not been increased since 1808, and their social status was threatened, while the amalgamation of departments and growth of yard accounts had greatly increased their work. More important, "the changes from wood to Iron Shipbuilding, and the numerous changes in the construction, armament, and the means of propulsion of ships of war" had created greater and much more varied responsibilities and entailed upon the constructors "a large and increasing amount of study and application". They petitioned the Board of Admiralty and their case was argued in the House of Commons in 1873; they were made to wait until 1875, however, for their rise.[36] Salaries now for the first time were graduated, rising in the case of the chief constructors at Portsmouth and Devonport from £700 to £850 per annum in annual increments of £25 (they had previously been £700 and £650 respectively).[37]

The pay rises did nothing, however, to alleviate the problem of work discipline. It was "a common topic of conversation, and...evident to anyone going about the Yard and Ships," Gordon Miller of the Transport Department reported in 1878, that a great deal of time was wasted. The Admiralty committees on the Professional Officers (1883) and Incidental Expenditure (1885) made similar observations. The problem was "attributed to lack of supervision by the superior professional officers downwards," and could be avoided, Miller thought, if the officers were not hampered by so much office work. Thus he recommended (although unsuccessfully) the appointment of civil assistants to provide the chief constructors and chief engineers with some relief. The Committee on the Professional Officers, like the Committee on Dockyard Economy in 1859, attributed weak work discipline to the leading men not being "sufficiently removed from the men whom they supervise to ensure efficient control." The solution, proposed by the Committee on the Professional Officers and adopted in 1884 (over the objections of the professional officers themselves, who had not been consulted), was to designate the leading men as inspectors (a nominal change) and to assimilate them more to management (it was hoped) by making them salaried officials, increasing their pay, and excluding them from piecework earnings and overtime pay.[38]

The management of stores was a question sufficiently knotty to be referred to the Admiralty Committee on the Store Department. The Committee reported in April 1876, and the last of Childers's reforms were undone. C.M.L. McHardy, the Superintendent of Stores,[39] had been altogether content with his position so long as the Controller was on the Board of Admiralty but highly dissatisfied ever since. McHardy argued that in order to be effective he needed to have direct access to the Board and be in direct communication with all the other Admiralty departments; this, however, had not been the case since 1872. He argued that the Superintendent of Stores should once more be a principal officer and under a superintending lord. It was Stewart's opinion, however, that since the Controller was responsible for the administration of the dockyards in its entirety and no work could be performed "without the employment of material as well as labour,"the demands of efficiency and expedition made it imperative that the Superintendent of Stores should be "a part of his own staff..., rather than...an independent head of a separate department." Everyone who gave evidence (except McHardy) agreed with Stewart. Nevertheless, the Store Branch on the Committee's recommendation was separated from the Controller's department and the Superintendent was created Director of Stores on the same footing as the Controller.[40]

The store departments at Portsmouth and Chatham were reestablished at the same time. All the superintendents who had served at those yards since 1870 (and Robinson also) had formed a poor opinion of Childers's scheme. The Committee concluded that the chief constructor's professional duties and lack of clerical and accounting experience made it impossible for him to discharge storekeeping duties effectively and that control over stores had as a result become a serious problem. The evidence concerning Devonport, however, was that "no arrangement for storekeeping could work better". Thus independent storekeepers were reappointed and once again made principal officers at Portsmouth and Chatham. The Committee did think, however, that "the Storekeeper should act in concert with, and defer as far as possible to, the Chief Constructor, in all important matters relating to the building department," although not that any formal instructions to that effect were necessary. [41]

Childers's bold experiment had been pronounced a failure and wound up. Annexing the office of chief engineer to that of master shipwright was unquestionably an ill-considered measure; separating the two again, but not providing some efficient means of coordination, was, however, to revert to an arrangement the deficiencies of which had been evident for over twenty-five years and presently became even more pronounced. The chief constructor had almost sole control over the ship up to the time of launching; it was afterwards, when the chief engineer installed the engines and other equipment, that close coordination between the two officials were imperative. Just as in the past, however, relations between these two independent, and mutually jealous, officers were more often disputatious than harmonious, especially since work at the stage after launching rapidly became more technical and involved after 1875. Both were partly responsible for installing armament, capstan and windlass fittings, steering gear, electrical and lighting arrangements, and boat hoisting gear; however, the point at which the responsibility of the one ended and that of the other began was not clearly defined. The work performed by the two departments consequently became so mixed up that it was not possible to establish direct responsibility for much of it, and economy of performance suffered accordingly.[42]

There was in the House of Commons, however, an influential body of opinion favorable to radical reform.   Charles Seely (chairman of the Select Committee on Admiralty Monies which had given Robinson palpitations a few years earlier) in 1873 asserted that "this mode of attempting to manage business by written instructions, instead of having efficient managers in the different dockyards has failed".   Thomas Brassey at the same time argued the necessity of appointing general managers "with plenary powers to carry out, according to their own judgment, the instructions received from the Admiralty". The Admiralty Committee on Incidental Expenditure in 1885 quoted with approval the view of the superintendent of Portsmouth yard that "the dockyards may be most efficiently and most cheaply worked" only with a civilian general manager.   Childers's successors, however, held stubbornly to the old organizational system and to naval superintendence, even though the superintendents usually served for no more than three years, whereas it took them one or two years to learn the job.[43]

The dockyard orders and instructions were still in the way they were written and recorded as much a source of confusion an impediment to good management as a century before. Not only were they very often imprecise and unclear, but an accurate record of them nowhere existed.

Instructions and orders were given from time to time and certain notations were made both at the Admiralty and in the Dockyards — sometimes in the margins of the printed books [of general regulations], but mainly in a

'Standing order book'. The Instructions given altered definitely perhaps one portion of a Para[graph] in the book, or a previous order, but questions would arise whether other parts would not also be affected or altered. Hence there was...great difficulty in ascertaining what the precise regulations were upon any particular point at a particular date.

An opportunity to do something about the problem presented itself when the dockyard regulations were revised in 1875. Orders after 1876 were issued in the form of printed circulars only (the premiss being that this would reduce their number and frequency) and printed on slips of paper which, whenever appropriate, were to be inserted in the General Order Book. This tactic failed, however; orders continued to proliferate and, despite admonitions, were still imprecise and unclear. When the dockyard regulations were revised again in 1882 (Brassey was now Civil Lord),

it was decided to eliminate...all Instructions relative to the mode in which the details of the various duties assigned to the Principal Officers and others are to be carried out, and to embody such details in Separate Handbooks; thus confining the General Book to the fixed principles on which the Administration of the Dockyards is to be carried out, and which is not liable to change.

Henceforth, supplementary orders were only issued quarterly, except for some urgent reason, and a committee edited everything for clarity and precision.[44]

Critics like Brassey had rather more in mind, however, when they spoke of instructions. Brassey wrote in a letter to *The Times* in 1872 that:

Every day's work in a dockyard depends more or less upon the directions received by the morning's mail from Whitehall. Hence it is that the correspondence at the Admiralty has become so voluminous, that important papers are forgotten and innumerable letters signed by persons who are totally ignorant of their contents.[45]

Just as in the eighteenth century, not even the smallest expenditure could be made without Admiralty authorization. Childers took the small step of empowering the superintendents to authorize additions and alterations without which a ship might be considered to be defective, but only up to the trifling sum of £50. Estimate for repairs —1325 from

Portsmouth alone in 1881[46] — generated mountains of paperwork yet were in many cases sent (as we have seen) long after the work had begun, or even when it was finished. The chief constructor at Portsmouth sensibly proposed in 1882 that repairs up to £500 should be dealt with locally. It was unrealistic, however, to think that the Admiralty would take such a leap (or that the Treasury would let it do so). William White, Barnaby's chief assistant, suggested £200, but knowing that amount would not go down either ventured to say that even £100 would save a great deal of work. This figure the Admiralty eventually accepted, in 1886.[47]

It was Gordon Miller who in 1878 demolished the notion that the expense accounts were of any practical use either to the yards or to the Admiralty. So much painstaking attention had been lavished on them that they were thought to have been made perfect, when in fact they had only become "cumbrous and unwieldy". They should have been for the chief constructors "a good check and a guide," whereas they were useless for either purpose, Miller said. All the chief accounts seemed "to be kept and prepared in such a way as to render...them of little or no practical value." The large excesses of expenditure over the amounts authorized by the Admiralty (as frequent now as ever before) were not clerical errors committed by the yard accountants, as was generally thought, but typically "errors of [accounting] principle". These observations led nowhere at the time, however. It was rather in the Store Department that the important accounting changes occurred in this period.[48]

Store management under McHardy underwent a sweeping reorganization after 1868. One of Childers's first acts as First Lord was to appoint a Committee on the Store and Purchase Departments. The storekeepers on the Committee's recommendation were relieved in 1869 of (the merely nominal) personal and financial liability for the stores in their charge; thus the bond which they had been obliged to give and the disruptive and expensive general survey to which they had been entitled when they retired (and upon which all insisted) were abolished. The continuous survey of stores, instituted in 1849 but indifferently observed (on the plea that the triennial surveys instituted at the same time made it unnecessary), was handed over to the yard accountants. Various irregularities were corrected; the store ledgers were sent annually (rather than quarterly) to London; and the number of forms and returns was reduced from 92 to 65.[49]

McHardy in 1870 substituted for the old fixed establishments (dating from 1844 and long since obsolete) variable, self-adjusting store establishments based on the preceding three years' consumption. Since the information needed for accurate store estimates was only available for about seventy of the principal articles, store estimates were henceforth based on the value of purchases during the same three-year period without regard to fluctuations of inventory. Information about current inventory and average annual expenditure was obtained by arranging the various articles (numbering around 20,000) under 120 heads and totalling them as money values using the prices in the Admiralty rate-book (periodically adjusted to reflect changes in market value). These changes naturally increased the work of the leading men, since they were now required to requisition stores under 120 heads rather than only three (general stores, rope and timber).[50]

Inventory control was much improved. Because the master shipwrights still did not know what quantities of stores they were to receive or how much money the parliamentary Estimates provided for store expenditure, they were inclined to stay on the safe side and overestimated their needs. The Admiralty thus was no less than ever in the dark about how much the yards actually required and when financial constraints made it necessary to trim demands did so without knowing how production might be affected. Beginning in 1870, however, each yard was informed as soon as possible as to just how much money would be available for the financial year and beginning in 1871 was for the first time given an outline of the year's naval program in time to use it as a guide in estimating what quantities of stores would be needed. Beginning in 1872, the Admiralty prepared separate estimates, thus making it possible to identify surplus stores and dispose of them as was appropriate.[51]

In 1873 a single annual demand for general-use stores was substituted for quarterly demands, and occasional demands were no longer permitted. It was now possible to substitute annual for standing contracts and to deal directly with the manufacturers themselves rather than with middlemen. Stores were consequently obtained more promptly and delivered more punctually, prices were lower, and the quality was better. Special-use articles like iron and armor plate were treated differently, however, and procured either by annual contracts, open-ended as to quantity and specifications, or by special contracts, under which they were ordered and supplied only as necessary. Such articles were exceptional also in being placed in the custody of the chief constructor immediately upon delivery.[52]

Iron and (after 1875) steel shipbuilding produced an explosion in the number of articles that needed to be stocked. Wooden ships had long since reached their furthest development, and the articles used in their construction had changed but little over time. Now, however, naval architecture and construction were evolving very rapidly; ships were of a much greater (and ever-increasing) variety and being furnished with more and more machinery all the time. New or improved articles were constantly being adopted, yet articles that were obsolete continued automatically to be stocked and even replenished. This runaway growth of stores was finally reversed in 1873, by adopting a continuously revised authorized list. At the same time, every article was given a distinct number for purposes of identification and control. The number of articles in inventory — 25,000 in 1873 — had by 1876 been reduced to between 14,000 and 15,000, and although the number rose over the next few years to 20,000, the value fell from £4.5 million in 1870 to around £2 million in 1882. Yet when the international crises of 1876 (in the Near East) and 1882 (in Egypt) occurred there was, because of better management, no shortage of stores as there had been in 1870.[53]

The metal mills were phased out after 1883. Production, which often exceeded demand, had been brought under control by 1877; nevertheless, the mills were a high-cost operation; it was cheaper to purchase their products from private suppliers, while the general use of steel after 1875 marginalized the iron mills.[54]

The physical disposition of stores contributed to poor management and consequent inconvenience to the construction department. It was of little use to arrange stores even in the best order unless they were arranged systematically as well. McHardy in 1882 recounted how

> formerly, articles of the same description were not only stored in different parts of the same buildings, but frequently in different buildings and in charge of different [storehouse] men, and consequently it was impossible to hold anyone directly responsible for the proper receipt and issue of stores.

This he had gradually corrected, however, as old inventory was depleted and new inventory was received, until the entire inventory of each article was stored in a single location, and the articles were grouped together according to the various categories in the store accounts, while each storehouseman was given charge of but a single class of stores. McHardy's systematic rearrangement made for better inventory control and simplified inventory-taking, reduced the work of the storehouse staff, and clearly defined each storehouseman's area of responsibility and enabled him to acquire expert knowledge of the stores entrusted to his care (better training was provided also). Unfortunately, the present-use storeholders were sacrificed to the new system of control.[55]

McHardy doubted that the continuous store survey by the yard accountant's staff was very effective; he did not like the randomness of it and suspected that it was not very thorough and, indeed, that it actually might not be conducted at all. He therefore introduced between 1880 and 1882 a number of compensatory measures[56] the most important of which was to have the accountant's staff make a quarterly comparison of the storehousemen's and the storekeeper's receipts and issues ledgers (the headings were expanded accordingly) and investigate all discrepancies. This was not to go very far, however, in dealing with the objection, raised by Miller already 1878, that what was called an audit of the store ledgers was "a mere check of details or comparison of figures."[57]

Storekeeping now involved much heavier and more exacting responsibilities than before, and in 1880 a higher standard of appointment was adopted. Storehouse boys were selected for sharpness and intelligence and received a thorough practical training, while advancement through the various grades (storehouseman, leading man, foreman) was by examination, and wages and salaries were raised.[58]

Rising doubts about the dockyards' performance led Northbrook in 1884 to appoint a committee to compare the cost of building and repairing warships in government and private yards (chaired by the President of the Institute of Naval Architects, the Earl of Ravensworth, and consisting of men identified with commercial shipbuilding).[59] The Committee reported that it was more expensive to repair and refit in private yards, and undesirable to do so in any case because of the highly uncertain extent of the work upon survey and the complicated question of how much should be paid for it, but that it cost less, and was more expeditious, to build in private yards and that the quality of the work, under Admiralty inspection, was in no way inferior. The Committee recommended, therefore, that the Admiralty should build more ships in private yards and that the dockyards should be confined more largely to repairs and maintenance and build only so many ships as to "ensure constant employment and perfect familiarity...with the latest details of the modern ship of war."[60] The Ravensworth Committee was not disinterested,[61] however, and its report should be viewed in the light of exceptional circumstances.

The exceedingly rapid evolution of materiel made the decade after 1876 (the year steel shipbuilding began) highly unusual. It was Admiralty policy to incorporate the latest materiel as it became available while a ship was being built, and this caused delays, involved the loss of materials and labor, and increased the cost. Seven or eight years were normal for building a battleship, and cost overruns (up to 45 per cent) were the rule. As a result, the building program was chronically underfunded, a situation that was exacerbated by the requirement that unspent balances (increased by delays) had to be returned to the Treasury at the end of the financial year. Contract-built ships, however, were largely exempt from these negative factors, for few alterations were ever ordered during construction, quite simply because the contractor presented his bill before performing the work, a deterrent not present in the case of dockyard work.[62] The situation in these years can be understood best by turning to the account by William White.

> During these ten years many important changes had to be carried out in Naval war *materiel*. Steel was first introduced as a substitute for iron in the hulls; steel-faced armour began to be used instead of soft iron; breech-loading guns were reverted to instead of muzzle-loaders, and the developments [sic] of quick-firing guns began. Hydraulic gun-mountings were in their infancy; the Vavasseur system of transferable gun-mountings first began to find favour; steam pressures were rapidly increased, and the earliest use of forced-draught in boiler-rooms of Her Majesty's Ships had to be experimented with. All these and other changes made rapid progress practically impossible and added to cost in a manner impossible to estimate.
>
> Ships were kept waiting for long periods while experiments were in progress to decide what kind of armour should be fitted. Others first intended to carry muzzle-loading guns eventually had breech-loaders. Others again were laid down and far advanced before their armaments were determined. The gun-designs were worked out after ships were begun; experimental guns, powder-charges and projectiles were tried and modified. Magazines were built, then altered and sometimes practically re-constructed. Gun-mountings had to be modified for guns. Quick-firing guns were added at a late period of construction. Electric lighting, both search and internal, had to be provided for although not contemplated originally in the designs. In short, delays and alterations of the most distressing character had to be accepted; and cheap or rapid construction became impossible.[63]

Continuous alterations and repeated interruptions, compounded by frequent diversions of labor to other work for which no provision had originally been made (e.g. commissioning, maintenance, repairs), devastated the morale of the yard officers. This comes through very clearly in the evidence the officers gave before the Committees on Dockyard Expenditure in 1885. Nor, apparently, was the morale of the workmen unaffected either.[64]

The Admiralty when ordering a ship from the dockyards initially provided incomplete — indeed, often only "the very baldest" — information about the hull and machinery. The rest followed as the work advanced, in contrast to the more complete specifications and drawings with which contractors were perforce furnished.[65] Questions were always arising but were not always settled promptly, a situation, however, which seems to have been improving by 1885.[66]

The working drawings and many of the related calculations were provided by the yard drawing office, whereupon the work was referred to the constructor in charge of the ship and by him to the foreman. The draftsmen were generally ill-qualified, however; the drawings might be insufficiently detailed, and the calculations might be in error, obliging the foreman to spend a great deal of time, which otherwise would have been devoted to the ship, "making or checking calculations that would have been done by ordinary Draughtsmen in a good...Drawing Office [in a private yard]." Everything was then returned to the chief draftsman, by him to the constructor, and then to the chief constructor for final approval. It was a time-consuming process.[67]

Construction delays were the result also of the Admiralty's tardiness in ordering, and the contractors' tardiness in delivering, materials. The chief constructor might send his demands away as quickly as possible yet even so might be kept waiting "a long time, sometimes months," before they were approved. This was because they were circulated in the Controller's office, mostly for informational purposes, before being forwarded to the contract department. Armor plates were a special problem; the dockyards, the Director of Naval Construction said in 1885, were "practically in the hands of two firms" which were suspected of having a secret agreement about how much they would supply the dockyards and of filling the orders of private firms first.[68]

The dockyard constructive staff were not equal to the managerial tasks confronting them. The appointment of the Surveyor of Dockyards, Frederick Barnes, was expected to change this, but the anticipated benefits were not forthcoming. Barnes himself was widely seen to be part of the problem. He visited the yards not frequently (as he was

instructed to do), but only occasionally, and then only concerned himself
with technical problems of construction; there was never enough time for
the officers to discuss larger questions of production and management;
everything was made the subject of correspondence, thus overburdening
the Surveyor's small London staff and causing production delays in the
yards. The Surveyor's assistants visited the yards more often but usually
confined themselves to the particular matter on which they had been sent
and in any case carried too little weight to do more. Barnes defended
himself and complained that he was trapped in an unwieldy bureaucratic
system. His responsibility, he told the Committee on Incidental
Expenditure in 1885, did not extend beyond making recommendations
and giving advice to the Controller through the Director of Naval
Construction (his nominal superior); such was the "complicated clerical
and Parliamentary system through which his proposals" had to pass
before they could be executed that he had little power to carry them out.
He was thought, however, to be insufficiently familiar with "the
requirements and business of the yards" and for that reason disinclined
to deal with management problems. Barnes does indeed seem to have
been out of his depth; his responses to some of the Committee's
questions were vague, and although he owed his appointment to having
for some years (1864-72) supervised the Admiralty's overseers of
contract ships, he himself had never built a ship or supervised labor.[69]

The deficiencies of the constructive staff became more and more
apparent as the 1870s wore on. The constructors in the dockyards had
a weak grasp of theory, while those at the Admiralty were unfamiliar
with the practical side of ship construction. Moreover, the academic
record of the shipwright apprentices at the Royal Naval College at
Greenwich (with which the School of Naval Architecture and Marine
Engineering was consolidated in 1873) was disappointing, and few of the
best students chose to remain in the dockyard service. This was
attributed by some to the failure fully to professionalize naval
construction; Great Britain, it was said, should create an elite corps of
naval constructors such as other leading naval powers (France, Italy,
Austria, Russia and the United States) had created already. The idea has
been attributed to White, and Brassey was agitating for it in 1879, but the
Committee on Dockyard Economy already in 1859 had proposed creating
a separate and distinct managerial class. Stewart was the driving force,
however; after leaving the Admiralty in 1881, he mounted a vigorous
campaign and persuaded Northbrook to appoint the Committee on the

Entry, Training, and Promotion of the Professional Officers of the Dockyards and in the Department of the Controller of the Navy. Brassey was chairman.[70] Thus was the Royal Corps of Naval Constructors, modelled on the French corps, created in 1883. It was an important event for the management of the dockyards and completed the professionalization of what had been, and to some extent still was, a trade.[71]

There was a vast gulf between the constructors at the Admiralty who alone were "experienced in designing ships and controlling the details of construction," and those in the dockyards who were called upon to build those same ships. The drawings and specifications sent down from the Admiralty were as a rule not very detailed; on the contrary, they were often quite sketchy and were, moreover, sent piecemeal. Indeed, Nathaniel Barnaby, the Director of Naval Construction, and his assistants disclaimed any responsibility for the details of a ship's structure; such matters, they argued, were for the chief constructor and the chief engineer to work out jointly.[72] Yet none of the dockyard constructors (not even the chief constructors) was "familiar with the process of ship design" or "competent to deal with those questions in an intelligent and scientific way;" their experience was limited to the "work incidental to building, fitting, and repairing ships," while the actual supervision of the work was in the hands of foremen who, their education having been confined to the apprentice schools, could only read a plan. Often clarifications, involving much correspondence and wasted time, were necessary, and Admiralty constructors had to be dispatched to the yards. Stewart argued that the two constructive staffs should be fully interchangeable (this was to be a function of the Royal Corps), thereby creating in each yard "a staff of naval architects capable of designing as well as building ships, so that projects for new ships would no longer proceed exclusively from the Admiralty staff and a beneficial competition would be established."[73]

The School of Naval Architecture had produced very disappointing results. Twenty-three students completed the course down to 1874 and were still alive in 1882; thirteen of these had received either a first- or a second-class certificate, but only seven were still in the service, compared with eight who had scraped through with a third-class certificate, which in the opinion of many should have disqualified them from becoming naval constructors at all. Moreover, the situation was getting worse. Twenty-one men completed the course between 1874 and 1882, one with a first-class, eight with a second-class, and twelve with

a third-class certificate. It was a record which compared most
unfavorably with that of the fifteen engineering students who graduated
from the Royal Naval College between 1874 and 1882, two with a first-
class and twelve with a second-class certificate.[74]

Stewart and White blamed the manner of recruiting, training and
promoting naval constructors. Young men were recruited not to a
professional career but as ordinary apprentices and served as such. Nor
were the few who went on to Greenwich accorded professional status
upon receiving their certificates. Even men with a first-class certificate
were only appointed supplementary dockyard draftsmen, positions equally
accessible to men who had attended the apprentice schools for four or
five years. The Greenwich men were obliged, moreover, after this initial
appointment, to rise through various temporary, non-professional
positions — foreman's assistant (supervising workmen), extra draftsman
at the Admiralty or assistant overseer of contract ships — for all of
which they were greatly overqualified. Permanent appointments, when
they occurred, were awarded by competitive examinations for which non-
Greenwich men were eligible also, and an education at Greenwich
conferred no special advantage. As a result, it was argued, many
Greenwich students, knowing how little use or advantage their education
would be for many years to come, fell into "a state of languor", and
many of the best afterwards left the service for greener pastures (most
often Lloyd's Register) as soon as they could. [75]

The argument in favor of a constructive corps had a distinct social
dimension that is revealing of British class-consciousness and class-
attitudes. Stewart explained that:

> the sons of persons of good social position...[are not] attracted into the
> shipbuilding department of the Admiralty. Only sons of working men or
> of persons in necessitous circumstances will submit to the severe
> competition, the rough life, and ultimate uncertainties of promotion.

The composition of the apprentice schools, indeed, was mostly working-
class with a lower-middle-class admixture. White pointed out that "not
one officer of the rank of constructor or chief constructor...has in recent
years so placed his son." If, he said, the sons of the Admiralty's leading
constructors were ever to follow in their fathers' footsteps, it would have
to be in a manner acceptable to men of that class. [76]

The engineer students, on the other hand, upon leaving the Royal
Naval College were immediately accorded professional positions and

status (most went to sea), and a good certificate was a highly important factor in promotion. This was assumed to be why they outperformed the apprentices at Greenwich, even though their social origins were either the same or only a little higher. The creation recently of a Corps of Naval Engineers and at Devonport of a training school (later the Royal Naval Engineering College), to prepare students for the Royal Naval College, had completed the process of professionalization with the result, it was pointed out, that the engineer students were now of a superior social standing and, of course, educational background. The failure to place careers in naval construction on a fully professional footing was clearly a serious lapse in an age that had already seen the enhancement socially of the old, and the proliferation of new, professions.[77]

It was thought that new life would be breathed into the management of the dockyards (and the social eligibility of a managerial career enhanced) if students of naval architecture were recruited from the engineer students and spared the rigors of an apprenticeship. Thus two engineer students upon completing their five years at Devonport might annually be sent to study naval architecture instead at the Royal Naval College, and scholarships were established to encourage private students (the response to the scholarships, however, was most disappointing). The Greenwich apprentices now spent a year at Devonport first; so closely, however, were naval architecture and marine engineering related that they were already receiving instruction in both.[78]

Greenwich graduates with a first-class certificate joined the Royal Corps of Naval Constructors as assistant constructors second class, either in the yards (either as professional secretaries to the chief constructors or general assistants to the superior officers) or at the Admiralty, while those with a second-class certificate became assistant constructors third class and general assistants to the superior officers. Men obtaining a third-class certificate were deemed to have failed and continued to be appointed dockyard draftsmen. It was important, however, to preserve the career prospects and morale of the foremen of the yard who were already in place; they might become assistant constructors second-class if they were recommended and passed a qualifying examination, as fourteen did between 1887 and 1901 (none was promoted afterwards). Promotion from the third to the second class was by seniority and to the higher grades by recommendation. Although it was expected that all assistant constructors would spend at least two years at the Admiralty, very few did, and interchangeability of staff was not achieved.[79]

Table 2.   Royal Corps of Naval Constructors, 1883.[80]

Admiralty	Dockyards
Director of Naval Construction	
Surveyor of Dockyards	
2 chief constructors	6 chief constructors (incl. one at Malta)
3 constructors	10 constructors (incl. one each at Hong Kong and Bermuda)
5 assistant constructors	
1 assistant constructor first class*	10 assistant constructors first class*
6 assistant constructors second class*	18 assistant constructors second class*
	9 assistant consructors third class*

* Indicates new position

The dockyards seemed in December 1868 to have reached a watershed.   Robinson had already outlined a revolution in organization and management as radical as any Bentham could have desired, and Childers now proved that it was possible to conquer hidebound bureaucracy. Their object was to bring the cost of production down, and with this in mind the organizational structure was revolutionized. The revolution failed because Childers had a nervous breakdown and then resigned after only twenty-seven months in office and before the revolution could be consolidated.   Counterrevolution followed but dragged on for years, a paradigm of the inertia that Childers and Robinson had hoped to overcome. To be sure, it was a mistake to annex the engineering and storekeeping departments to that of the master shipwright; the arrangement was only experimental, however, and Childers might have been persuaded to adopt Robinson's sensible proposal to place the three departments under a general manager. The wisdom of making the Controller a lord of Admiralty was eventually vindicated. The subordination of the Storekeeper General's department as a branch of the Controller's department worked very well at first; McHardy's complaint was that he no longer had direct access to the Board of Admiralty and the other Admiralty departments after 1872.

Goschen wisely separated the constructive and engineering branches of the Controller's department once again. His chief contribution, however, was to take the managment of the yards out of the hands of the Chief Constructor/Director of Naval Construction and appoint the Surveyor of Dockyards, even though the improvements in the economy and efficiency of operations which this measure was expected to produce still had not been realized in 1885.   The Royal Corps of Naval Constructors completed the professionalization of naval construction and enhanced prospects for better management.   Yet the suffocating system of centralized management by correspondence was in no way relaxed. The chief constructors continued to be so loaded down with non-professional paperwork (mostly accounts) as to be left with too little time for enforcing work-discipline, which remained as great a problem as ever. There was, moreover, no effective means of controlling the cost of production, or even knowing just what the cost was, despite the number of accounts the compilation of which consumed so much time. McHardy reorganized the management of stores, revamped the store accounts and improved inventory control, all of which represented a considerable achievement, yet neither the store nor any other accounts were subjected to what even at the time was considered to be a true audit.

Such changes as occurred in organization and management between 1868 and 1885 yielded with few exceptions only indifferent results. Indeed, the same challenges faced the dockyards at the end as at the beginning of the period, and private yards still built for the Admiralty at less cost and in less time than the royal yards.

Chapter 8

The Watershed, 1885-1900

It was Lord George Hamilton who brought the dockyards to the watershed whilst he was at the helm of the Admiralty from 1885 to 1892, except for the few months from February to August, 1886 when the Liberals were in office again and the Marquess of Ripon was First Lord.

Hamilton came to the Admiralty at yet another awkward juncture. W.T. Steed in September 1884 had published in *The Pall Mall Gazette* a series of electrifying articles, "The Truth about the Navy", which shattered years of complacency and forced Northbrook to add £3.1 millions to the naval program. The Ravenswood Committee reported in October, however, that the dockyards were not really fit to build ships and should largely be restricted to repairs and maintenance. Northbrook, nonetheless, failed to come to grips with the imperative of how to build faster and cheaper in the dockyards and make them competitive with private yards.

Hamilton was determined to prove the Ravenswood Committee wrong and that the dockyards could be fully competitive with the so-called "battleship" yards. Unlike Childers, he was a stranger to naval administration and had no preconceived ideas (an advantage, as it turned out); moreover, his style of management, unlike Childers's, was collegial. He appointed, and took the advice of, four committees of inquiry into Admiralty and dockyard expenditure and administration — indirect dockyard expenditure[1] (chaired by Vice Admiral William Graham, the Controller of the Navy, 1885-88),[2] financial arrangements at the Admiralty, the Accountant General's department (both chaired by

C.T. Ritchie, President of the Local Government Board), and purchase and contract (Arthur Forwood, Parliamentary and Financial Secretary) — which reported between September 1885 and February 1887. Thus Hamilton's reforms, having been so widely canvassed before they were undertaken, were unencumbered by the controversy and confusion that had helped to sink Childers's reforms. Years later Hamilton claimed, with no false modesty, to have "so revolutionized the whole system of administration" as to make "the Royal dockyards...the cheapest and most expeditious building yards in the whole world".[3]

The revolution coincided with the first stages of the naval arms race that culminated in the First World War and thus came none too soon. Fear of the combined naval strength of France and Russia led to the Naval Defence Act of 1889 and Hamilton's unambiguous enunciation of the two-power standard.[4] A further scare from the same quarter led to Earl Spencer's[5] even more ambitious program of 1895, and expenditure, which since 1889 had averaged £5,017,208 per annum, rose in the next five years to an average of £9,204,548.[6]

Graham's Committee reported that the dockyards were indeed uneconomic and inefficient and that the solution was to model their management as much as possible on the management of private yards. The construction and engineering departments were uncoordinated, accounts duplicative, the clerks overworked, the returns unnecessarily voluminous, and storekeeping arrangements inconvenient; audit was defective, financial control non-existent, and control over labor ineffective. The Committee specifically recommended:

1. strengthening Admiralty executive and financial control;
2. appointing professional managers to assist the naval superintendents at Portsmouth, Devonport and Chatham;
3. relieving the professional officers of non-professional duties;
4. continuous supervision of labor and work by "competent subordinate officers...when the officers immediately in charge are otherwise engaged;"
5. improving storekeeping arrangements; and
6. introducing a system of independent audit.

These recommendations obviously involved spending more money, not less; however, the Committee, repudiating the cheeseparing and denouncing the false economies of the past, argued that the cost would be a mere trifle compared to the enormous waste that would certainly be eliminated.[7]

The reforms that followed started at the top and affected every level of management and supervision.

The evidence given before Graham's Committee exposed the deficiencies of the dockyard branch and of the Surveyor of Dockyards himself. The Committee recommended strengthening the hand of the Surveyor of Dockyards so as to ensure "competent, systematic, and businesslike supervision over the...dockyards," and Ritchie's Committee suggested that the Admiralty might look outside the dockyard service for Barnes's successor.[8] Barnes retired early in 1886, and Ripon appointed as his successor Francis Elgar, with the new title of Director of Dockyards. The Controller was at the same time restored to the Board of Admiralty.

Elgar was a superb choice. He had gone to the Admiralty, and was there befriended by Reed, following graduation from the School of Naval Architecture in 1867. When Reed resigned and went to manage Earle's yard at Hull, Elgar went with him as professional assistant and a few years later, when not yet thirty, succeeded him as resident manager. In 1879 he became adviser on naval architecture to the Japanese government and in 1881 set up as a consultant. When the first professorship of naval architecture and marine engineering in Great Britain was established at Glasgow University in 1883, Elgar was chosen to fill it.

Elgar, as Director of Dockyards, was (unlike Barnes) head of a separate and distinct branch and equal in rank to the Director of Naval Construction. His functions were the same as those of his predecessor, but the revised instructions (issued by Ripon) strongly emphasized the importance of visiting the dockyards frequently and consulting with the superintendents and constructors on technical and management problems. The significant difference was that the Director was directly and personally responsible to the Controller and possessed the requisite "discretion and freedom of action" for ensuring that dockyard work was properly and economically executed. The new branch, however, was responsible for its own administrative work and this, since the volume of work was growing, seriously encroached on the time the Director was able to spend personally assisting the yard constructors, a difficulty that was dealt with in part by appointing naval constructors to assist the superintendents at Portsmouth, Devonport and Chatham.[9]

Elgar's instructions created a contretemps with William White, the new Director of Naval Construction appointed by Hamilton. White had been Elgar's contemporary at the School of Naval Architecture and was the greatest naval architect of his day. He left the Admiralty in 1883 to organize and manage Sir William Armstrong's new warship yard at Elswick and to design warships for foreign and colonial navies. It was a great coup, therefore, to have brought him back to Whitehall. White was of the opinion, however, that the Director of Dockyards should generally be subordinate to the Director of Naval Construction[10] a relationship, needless to say, which a man of Elgar's prominence would have found intolerable. There was another aspect also. That was Elgar's friendship with White's fiercest critic, Sir Edward Reed, M. P. (as he now was). Reed was a Lord of the Treasury and may have been instrumental in Elgar's appointment.[11]

White fancied a more active role in dockyard operations than his predecessor Sir Nathaniel Barnaby (Reed's brother-in-law) had played. Egged on by the partisan Graham, he fired off a vigorous protest to Ripon. Ripon, however, was not to be swayed. It was his view that the 1872 instructions left "the position of the Director of Dockyards doubtful and precarious" and would be fatal to the new policy that the dockyards should be treated the same as the Admiralty treated private contractors and be required to complete work on schedule and without unauthorized cost overruns. This was Hamilton's view also. Still, White's self-regard had to be soothed, and he was accorded the additional, although merely honorific, title of Assistant Controller.[12]

There was inevitably a good deal of "friction and unpleasantness" between White and Elgar, although relations appear to have improved when Graham resigned and was succeeded by Rear Admiral John Hopkins in 1888. White again tried to get his way when Elgar resigned and joined the Fairfiled Shipbuilding and Engineering Company at the beginning of 1892. Hamilton, however, thought that things had worked out quite well just as they were.[13]

Elgar's successor, James Williamson (1892-1906), had attended the Portsmouth apprentice school for five years but was prevented from obtaining a higher education by the closing of the second School of Naval Architecture in 1853 (he himself was fourteen at the time). He was an assistant overseer of contract ships when in 1865 he left the service for Lloyd's Register, where he eventually became principal surveyor. In 1881 he joined the big Clyde shipbuilding and repairing firm of Barclay, Curle and Company as managing director. Thus had practical experience equipped Williamson well for the position he now undertook, even though Barclay, Curle did not build warships.

Graham's Committee called the absence of coordination between the construction and engineering departments "one of the most serious blots on dockyard management", giving rise to friction and creating unnecessary expenditure. There was no one in a position, "analogous to that of the proprietor or manager of a private concern, upon whom the efficient conduct of the business would generally devolve." The superintendent was "merely the channel of communication between the Admiralty and...[the principal] officers;" he possessed "no distinct or direct responsibility," exercised no personal or independent control over officers and workmen, and, having insufficient technical knowledge, was obliged to accept the judgment of the professional officers as to whether the work was being performed economically and efficiently. A secretary (a dockyard clerk whose job it was to process routine paperwork) constituted his entire staff. He was perforce office-bound and disliked sending for the professional officers and taking them away from their work. The Committee likened the superintendent to a general without a staff and deplored his isolation from the professional officers whom he supposedly commanded.[14]

The Committee took the traditional view that because of "the constant intercommunication which must take place between the naval and dockyard officers," the existing system of naval superintendence was necessary, even though it had failed. It would be found to work, however, if just two things were done: endowing the superintendent with sufficient authority and responsibility and, even more important, providing him with the assistance of a naval constructor whose sole responsibility would be to coordinate and control production. Thus were civil (later civil technical) assistants to the superintendents of the three principal yards appointed in 1886.[15]

The civil assistant advised the superintendent "on all questions of a professional or technical character," coordinated the departments, acted on behalf of the superintendent "and with his authority, but under his directions, in all matters connected with the Civil Departments of the Yard," and was responsible for the distribution of labor and the appropriation of materials and for ensuring that the work was performed satisfactorily and economically. It was not intended that he should be a general manager, even though his powers were broad, but rather a kind of executive assistant who would make it possible for the superintendent to function as a real, rather than as a figurehead, general manager. There was, however, a certain ambiguity about the arrangement, enough to enable the civil assistants little by little to usurp the executive and financial responsibilities of the chief constructors and chief engineers and become as it were through the back door "practically...separate officers [i. e. quasi-general managers] intermediate between the Superintendents and responsible Heads of Departments." This eventually raised concerns about diminishment of the professional officers' sense of responsibility and the effect on the management of their departments. Did the civil assistants improve management, though? They were never really evaluated in that context, and the office was abolished in 1905. The superintendents under their influence do seem, however, generally to have assumed a more active role, although not always for the better.[16]

Graham's Committee saw the officers' instructions not as an aid but rather as a barrier to good management, an attempt to adapt the management system of a bygone age to the complexities and radically different circumstances of a modern dockyard. The instructions deprived the officers of that discretion without which it was not possible to act responsibly and effectively, forced upon them positions which they could not occupy and responsibilities which did not exist, and conferred powers which they could not exercise. The Committee sensibly recommended that "while the instructions should indicate clearly the principles which should govern the administration of the Dockyards, they should allow a large measure of latitude to the Superintendent and his officers to enable them to control the business". Thus when the instructions were revised a few years later, they were "strictly confined to matters of principle", and the handbooks, introduced in 1883 for tighter Admiralty control, were abolished. This, however, was not quite the victory for decentralization as it might seem to be, since the Controller might issue his own handbook containing detailed and supplementary instructions.[17]

The war on the redundant returns and accounts that lumbered the officers had been going on fitfully, and in the long run not very successfully, ever since the 1780s. The problem had grown again and was now attacked even more vigorously. Many returns and accounts, instituted in a bygone age, a fast, efficient postal service and the electric telegraph had long since rendered unnecessary, while most were of no practical use anyway. Eighty-two returns and accounts were either abolished or revised on the recommendation of Graham's Committee, and the officers were admonished to do their part in reducing paperwork even more. The weekly progress of works, of which, as we have seen, so little use was made (but costing £500 per annum just to compile, plus the expense of abstracting it in London), was abolished and no longer encroached upon the time of assistant constructors, foremen, inspectors and leading men. The diaries of work, consisting of around 23,000 items, were confined to building and extensive repairs, additions and alterations, and only "sufficient information to distinguish between the various sub-divisions of the work" was henceforth to be given. It was at this time that permission was given for local authorization of repairs up to £100 (recommended in 1882), another measure which saved a good deal of time and paper.[18]

The establishment of financial control, remarkable until now, in the words of Graham's Committee, for its "entire absence", was a very considerable achievement. The details of dockyard work (the basis of the expense accounts) were recorded in the case of labor by the leading men (whose veracity was not above suspicion) and in the case of materiel by the storekeeper. The yard accountant collected the details weekly but depended entirely on information provided by the professional officers to distribute the cost of labor over the various ships and incidental services, an exceedingly involved process (32,000 items of wages and stores were posted each week at Portsmouth). So long did it take to do all of this that, as we have seen, the officers only received the expense accounts long after they were of any use for controlling expenditure and for this reason felt "absolutely compelled" to keep their own (unauthorized) records of labor (and to some extent of materials) and to employ the necessary staff. The officers, indeed, considered the accountant to be "a drag and a hindrance [rather] than an assistance in carrying out work for which they were themselves entirely responsible".[19]

The recording of the work and the preparation of the expense accounts were transferred in 1887 to the newly created Expense Accounts Branch of the Controller's department.[20] The Inspector of Dockyard Accounts reviewed and examined expenditure, furnished full statistical and financial information to the professional officers, the Controller, the Accountant General and the Financial Secretary, and visited the yards and supervised the local staff. The functions of the local yard expense office were to:

1. record and measure the work and (after 1891) calculate piece wages,
2. prepare the expense accounts of ships and services and the cost accounts of ships under construction,
3. compare weekly actual and the estimated expenditure,
4. prepare the expense accounts of factories and workshops), and
5. prepare the stock valuation account (instituted in 1888) and make test comparisons of the quantity and value ledgers of stores.

The accountants Edwin Waterhouse and Richard Mills,[21] appointed by the Treasury in 1888 as a committee to report on the new system of dockyard expense accounts, were impressed with the great "practical value" of the simplified "method of record and account-keeping" that had been adopted. The accounts of the expenditure of wages and materiel were "so closely written up from day to day" that the officers were handed "before the end of every week a [detailed] statement of the expenditure of the preceding week". Even so, the professional officers continued until 1902 the long since forbidden practice of keeping their own records.[22]

Waterhouse and Mills assured the Treasury that the accounts were as trustworthy as possible. The recorders of work visited each man at his work twice daily and recorded the items on which he was employed and the length of time he spent on each; moreover, the recorders were independent of both the men and the officers, and their "districts" were changed periodically. The recorded times were converted at the expense accounts office into money equivalents (at the applicable wage rates), and the sums were posted to the accounts of the various ships, and so on; the total of day pay was then checked against the cashier's wages sheets (prepared from the timekeeper's independent record of actual attendance). Waterhouse and Mills could "suggest no simpler or better method for securing a trustworthy allocation of labor [costs]".

> The very multiplicity of the entries [in the accounts] and the number of persons employed on them...[who] are in no sense under the eye of the professional officers, and have no interest whatsoever in the results of the figures which they record, practically preclude the possibility of any improper dealing with the figures, such as intentionally charging to the cost of one vessel an expense which should be borne by another; while the method adopted provides that the clerical accuracy of the work is proved in the aggregate. No outside examination of these enormous details is therefore, in our opinion, necessary....

Nevertheless, the expense and manufacturing accounts were after 1888 subjected to a test audit by the Comptroller and Auditor General, thereby dealing, in these two cases, with the complaint of Graham's Committee that there was "in effect no audit at all."[23]

A stock valuation account, thought to be crucial to financial and cost control, was still missing, however. The account established in 1864 had proved to be of no practical use and was abolished in 1879. There was (since 1870) a record of the value of store issues, but at invoice (i. e. purchase) prices rather than rate-book (i. e. market) prices; thus this account also was of little practical use. Arthur Forwood, the Parliamentary and Financial Secretary, in 1888 established a new stock valuation account showing the value at rate-book prices of all store department transactions. The comparison annually of the balances of this account and the store accounts was substituted for the former examination of store accounts by the yard accountants. The new stock valuation account served several purposes: as an independent check on the store and expense accounts, verification of the storehousemen's and storekeeper's accounts, and the basis of an abstract of actual issues of stores under the various categories of yard expenditure, and provided information for preparing the estimates for the following financial year.[24] Forwood's stock valuation account was not all it was cracked up to be, though, and was discontinued twenty years later; the procedures necessary to compare an account of values and store accounts of quantities were simply too complex and cumbersome, and other accounts were found to be more useful in preparing the estimates.[25]

Independent verification by unannounced, random surveys of the balances of inventory shown in the storekeeper's ledgers was, in the words of Waterhouse and Mills, "the arch upon which any proper system of check upon store accounts must rest." A continuous, random survey was conducted by representatives of the storekeeper and accountant jointly until 1886. In that year, however, the Treasury intervened and required the substitution of independent inventory-takers audited by the Comptroller and Auditor General, thereby bringing dockyard practice in line with commercial practice. The survey was undertaken at first by the storekeeper's own assistants, a practice unacceptable to the Treasury and to Waterhouse and Mills, since it was "of the first importance that the arrangements for stocktaking are such as cannot be reasonably impugned or challenged." The survey in 1889, therefore, was transferred to inventory-takers from the Storekeeper General's office in London. This, however, did not satisfy the Public Accounts Committee's understanding of "independent", and in 1898 representatives of the Accountant General were associated with those of the Storekeeper General.[26]

Thus was created at last a system of accounts for providing the information necessary for financial control, but in a form highly useful for cost control also, a system which, when it was evaluated twenty years on, was deemed to have proved itself over time. Nonetheless, Ashworth's judgment that "the accounting practices of naval establishments were probably no worse than those prevailing in most private businesses...[but] because of the nature and scale of the operations" needed to be better, is fair.[27]

The developments of the late 1880s, along with the professional status conferred by membership of the Royal Corps of Naval Constructors, gave a tremendous lift to the officers' morale, which, it is clear from the evidence given before Graham's Committee, was pretty low in 1885. Forwood observed after visiting the yards in 1890 that

> A very commendable change is perceptible at the Royal Yards. A keen desire is evinced on the part of the officers to produce work at the lowest cost, and in this respect a spirit of emulation between yard and yard, and between the yards and contractors has sprung up.[28]

That spirit of emulation was above all owing to the new system of expense accounts, thus fulfilling C. G. Bedford's prediction thirty years before that only "a definite system for utilizing the...accounts of expenditure" was capable of overcoming the "apathy" that pervaded the dockyards.

Work discipline, however, remained a dark cloud which would not be dispelled.

Graham's Committee acknowledged that there were many industrious workers but found, nevertheless, that idleness was not only "practically unchecked" but increasing and that the men even posted look-outs so as not to be detected. The superior officers appeared "to be powerless...and the junior officers...to be either apathetic or too much in the hands of the men." Barnes, speaking from years of experience overseeing contract ships, told the Committee that idleness, far from being peculiar to the dockyards, was a universal problem. Indeed, at Armstrong's yard at Elswick the men for the first two hours of the day were mostly engaged in making preparations for their breakfast-break. Idleness was certainly significantly greater in the dockyards than in private yards, however; a mass of consistent evidence points to this conclusion. The Committee agreed with Barnaby that the problem could only be solved if the yard officers, like their counterparts in private industry, were given the power of dismissal. As it was, "the closest investigation, even magisterial in character, may follow upon a report against a man by his officer." The result, said Barnaby, was "a natural feeling amongst the officers that...it is sometimes desirable to gloss over or wink at idleness on account of the

difficulty and trouble and annoyance one has, to get the man punished. "
The power to dismiss was not a realistic proposition, however; even the
recommendation the Committee did make — that the officers should be
able to levy fines, impose reductions in grade or pay, or suspend pending
a hearing by the superintendent — the Admiralty did not accept.
Established workmen were still dismissed only for the gravest offenses,
and political pressure continued to make the dismissal of hired men (by
now 74 per cent of the workforce) extremely difficult. It was the same,
or even worse, in the French dockyards, as the British naval attaché in
Paris, Captain Alfred Paget, reported in 1897, referring to complaints
that it was impossible "to get a fair day's work out of a French
dockyardman, as the men are all serving permanently, sacrificing a
percentage of their salary for deferred pension and except in very flagrant
cases of shulking, cannot be dismissed [because of political influence]".[29]
Weak work-discipline was also a factor of weak supervision. The
foremen were strictly enjoined to a "constant presence with and vigilant
inspection of the workmen"; however, one foreman might oversee two
or three ships and 200-400 men, all at the same time, and thus had no
choice but to leave inspectors (the former leading men) in charge for
extended periods.[30] Graham could not understand how if there were 400
men, and 16 gangs, and all the inspectors were of equal rank, the work
could be performed in a systematic and businesslike manner. It could not
be, said Barnaby; "an officer of authority should constantly remain in
supervision of work going on board each ship where a number of men
are employed." The solution was to put one inspector in full charge
when the foreman was absent.[31]

The inspectors, however, were themselves accessories to the men's
idleness. Elgar at first dealt with this problem by requiring a
performance report on each inspector and limiting the annual salary
increment to those getting a favorable report. This scheme, like
Middleton's a century before, came to nothing, however; the inspectors
looked upon "their increments as a vested right", and the professional
officers, knowing that a bad report would "raise an amount of feeling and
dissatisfaction that would doubtless find its vent in political agitation,"
simply reported all as being either "excellent" or "very good". Elgar
next instructed the superintendents to report unfit inspectors; only one did
so, however.[32]

Evidence was accumulating that most inspectors, especially those
acting in place of foremen, were indeed unfit. Elgar in 1889 grasped the
nettle and began reducing their number by attrition and replacing them

with chargemen, non-salaried supervisors who could be required, when necessary, to work along with the men and were, in fact, exactly like the old leading men. By 1898 the number of inspectors had been reduced from 155 to 69, and the position of chargeman had become an apprenticeship for that of inspector; meanwhile, the examination for an inspectorship was made more rigorous (the standards for education and practical knowledge were raised). It was expected that the foremen and assistant constructors would personally keep the chargemen to the mark; this, however, was incompatible with their other duties (each was in charge of several ships or shops). At the urgent request of the dockyard officers, the supervisory structure created by Sir James Graham, and abolished by Childers, was essentially restored in 1899, each inspector being put over three chargemen and functioning rather like the original inspectors.[33]

Meanwhile, shirking and idling continued to increase. There was, however, a massive expansion of the workforce to 32,000 after 1895[34] (from 18,000 in 1890), and this undoubtedly overstretched the ability of existing systems to cope. Plainclothes police deployed on bicycles apprehended numbers of offenders, but the punishments which were meted out were too light to have any deterrent effect. Sir Charles Howard, late assistant commissioner of the Metropolitan Police (1890-1902), reported in 1903 that "the extent to which shirking is now carried on is an open scandal," and the Admiralty issued "stringent orders" on the subject, but the problem persisted anyway. It was said in the House of Commons in 1904 that many industrious men notwithstanding, the dockyards were a refuge for men who "did their work in a malingering way, and who did not do a great deal of work." No wonder Elgar complained that it was not possible to run "the Government dockyards in the same satisfactory way as private establishments".[35]

As important as Hamilton's innovations were, they nevertheless could not of themselves have met the objective of building faster and cheaper, although they helped to create the conditions for doing so. It took up to seven years, and five years and nine months on average, to build the eight battleships that were laid down in the dockyards between 1880 and 1884. Yet the French dockyards took just as long, and the one battleship that the Admiralty ordered from a private yard in the same period took six years and five months to build. Building was speeded up after 1885 by massing far greater numbers of workmen on each ship than previously and by maintaining full and continuous production from start to finish.

The officers at Portsmouth were dubious at first when told that the *Trafalgar*, the biggest battleship up to that time, was to be built in this manner and argued that "even if the work were kept fully manned all the time," the ship still could not be built in less than five years and that so many men could not be employed profitably. They were proved wrong; the *Trafalgar* was laid down in January 1886 and completed in March 1890. She could have been completed in three-and-a-half years, however, had the work not been held up by alterations and additions and by delays in obtaining materials and fittings. Indeed, Elgar in 1891 did not doubt that the dockyards "could now undertake to build any Battle Ship in 3 to 3½ [years] from the laying of the keel plate." Production time fell, however, in private yards also. The six battleships of the 1889 program that were built in the dockyards were on average completed in three years and nine months, while the average for the four built in the yards of the so-called "battleship" firms was a little less than three years and six months. Average building time fell markedly during the first part of the Spencer program to two years and seven months for the seven dockyard-built battleships and to three years for the two built in private yards.[36]

The new system of production reduced not only building time but cost as well. Figures compiled for Elgar showed a marked fall in the cost of labor between battleships built in the dockyards between 1875 and 1885 and between 1886 and 1891, although both he and White were careful to warn that the comparisons "were not absolutely accurate as between different types of ships" but were "only *indications of economies*". The cost of labor for the hulls and fittings of the thirteen turret battleships laid down in the earlier period was £42.8 — £57.4 per ton, the average being £51.2, whereas it was £36.3 for the *Trafalgar* and £38.9 for her sister ship the *Nile*. The *Nile*, however, took rather more than five years to build, thus apparently proving the point about rapid building being cheaper.[37] Ten of the earlier ships exceeded the estimated cost of labor by 3-45 per cent (five of them by more than 20 per cent), and three were under the estimate by 0.7-5 per cent; however, the *Trafalgar* cost 31 per cent, and the *Nile* 25 per cent, less than the estimate,[38] even though many expenses formerly counted as incidental expenses were charged direct to the ships themselves, the weight of the two ships was much greater than that of any laid down in 1875-85, and numerous "alterations and additions, some of which were very costly," were made.[39]

Continuous production was only possible so long as there was continuous funding. The Admiralty convinced the Treasury of the considerable financial saving in the long run, and as a result the Naval Defense Act of 1889 provided for the unspent funds for each ship in the program to be carried over from one financial year to the next, instead of being returned at the end of the year. Administrative problems led to disputes between the two departments, however, and questions were raised in Parliament; as a result, subsequent naval programs were funded in the old way. Nevertheless, the first part of the Spencer program was completed in even less time than the 1889 program. The battleships laid down in the dockyards between 1896 and 1901 took longer to complete, the average time being three years and five months (but still less than in the 1889 program), while the average for those built in private yards rose to three years and eleven months. The 1889 program, however, coincided with a shipbuilding boom, whereas the Spencer program coincided initially with the deep slump of 1892-95. The boom that followed lasted until 1901 and, along with the naval programs of the period, overtaxed the country's shipbuilding and manufacturing capacity. Meanwhile, the engineers struck nationally from July 1897 to February 1898. The strike, even though it directly affected only engineering works and the manufacture of armor plates, nonetheless disorganized and created bottlenecks in both the dockyards and private yards for several years afterwards; deliveries of materials were held up, payments to contractors were deferred, and consequently the building programs of these years never exceeded the authorized expenditure.[40]

How does Hamilton's boast that the dockyardmen were the most efficient shipbuilders in the world stand up? The facts, insofar as they can be known, may be allowed to speak for themselves. The time taken to build a battleship fell more markedly — and, more important, after 1893 was less — in the dockyards than in private yards. In Germany alone, among the other naval powers, were battleships built as fast as in Great Britain between 1888 and 1902, the average time from commencement to completion being about three years and four months in both countries, whereas in France it was five years and in the United States four years and four months. The battleships of the *Royal Sovereign* class (laid down in 1889-91) cost on average 6 per cent less in the dockyards than in private yards, those of the *Majestic* class (1893-95) almost the same, and those of the *Canopus* class (1896-98) 3 per cent more.[41] It is not possible, since there are no extant accounts, to reconstruct the actual costs incurred by the "battleship" firms; these firms, however, would put in very low tenders when trade was slack, so as to avoid making an even greater loss by keeping their plant idle.[42]

Meanwhile, the dockyards were re-equipped with the most up-to-date labor saving machinery and appliances and were fully abreast, if not in most respects ahead, of private shipyards. Still, the British lagged well behind their foreign competitors in mechanization; they had, however, a long head-start in iron and steel shipbuilding and, consequently, a much more efficient workforce.[43]

No longer was it necessary to confront and adapt a bewildering proliferation of new, insufficiently understood or tested, and rapidly evolving technical developments in steam power, armor and armament. By the late 1880s sufficient experience of these had been gained, the earlier problems had largely been worked out, and ship design was based on a practical settlement of armaments. Nevertheless, the appetite for incorporating the latest improvements in machinery, gun-mountings and fittings during construction (as a rule without having been approved by the Admiralty Board) was still not quenched, and production as a result was often held up. In 1902 the practice was sharply criticized in view of the intensity of the naval arms race; alterations, said the Admiralty Committee on Arrears of Shipbuilding, should be ordered only for "the very gravest reasons" and only with the Boards's approval.

> In view of the importance of keeping the Fleet at all times up to the strength which is considered necessary, it is better...to complete five or six battleships as rapidly as the resources of the country will allow, and to include any improved appliances...in a subsequent batch of ships.[44]

Ripon had earlier attacked another hindrance and issued an order that "the designs and specifications for ships should be supplied to the Dockyards in as complete a state as they are supplied to private shipbuilding firms." The order had little effect, though, so long as the status and pay (a time wage of 12s. per day) of the Admiralty drawing-room staff were insufficient to attract enough capable draftsmen. Not until 1890, though, were the staff put on salary and given a rise, to £250-300 per year, thereby achieving professional status.[45]

Piecework was revived, on White's recommendation, in 1887, although only for building.[46] The yard officers found fault with piecework, however, largely because the work of individuals could not be distinguished from that of the whole, thereby diluting the incentive value. Thus they devised tonnage payment, a highly complicated system in which the men were paid according to the weight of the material built "into the whole or a section of a ship". This was introduced experimentally in 1889 but soon found to be quite unsatisfactory (e. g. worker output did not increase very much) and abandoned. Piecework was criticized for the same reasons as in the past but was continued anyway, although on a restricted basis. Only about 17 per cent of Portsmouth's workforce was paid a piece wage in 1912, whereas most men in private yards (and about half of all French dockyardmen) were so paid.[47]

Greater expedition in the procurement and delivery of materials also speeded up production after 1885. The Director of Contracts now consulted more closely with the yards before making contracts, penalties were levied for non-fulfillment, and the yard storekeepers sent specifications directly to the contractors as soon as they were received from the officers rather than to the Admiralty first. Special articles (e.g. armor plates) continued to be ordered through the Controller's office; the specifications were immediately sent from thence to the contractor, however, and were only subsequently circulated, solely for informational purposes, in the office.[48] The present-use storeholders, abolished in 1882, unfortunately were not restored, despite the Committee's recommendation and the fact that the officers set them up anyway.[49]

The dockyardmen and the Admiralty confronted each other uneasily between 1889 and 1893, at the same time that industrial relations in Britain lurched into a period of rapid unionization, rising militancy, and turmoil. Unrest over a multitude of grievances, both great and small, boiled over and swept the entire workforce into a single movement that generated wide public sympathy and reverberated politically.[50] When the moment came in 1890 to receive the men's petitions, 224 (the most ever) were handed in. Not since the bloody strike of 1801 had discontent run so high. It did not come to a strike this time, although there were murmurings of one, and the naval program was not at risk, for, as the secretary of the Portsmouth branch of the Associated Shipwrights Society[51] said later, a strike against the government would have been futile.[52]

Wages were the most prominent grievance. They had not been increased since 1873, whereas the shipbuilding boom and recent work stoppages had everywhere else lifted shipyard wages to new highs. The question of dockyard wages was complicated, however, by the structure of the workforce. The shipwrights (now only 20 per cent of the dockyardmen) had long since become all-round workers in iron and steel, but the Admiralty had reneged on its promise to raise their wages accordingly and was paying them only 32s. per week, whereas men doing the same work in private yards, but belonging to various other trades, were paid on average 40s. per week. In fact, some of these same trades were employed in the dockyards also, at higher wages than the shipwrights, although still less than they would have gotten elsewhere.[53]

It was Elgar's policy as much as possible to substitute cheaper for more expensive labor on new construction, a policy that led to demarcation disputes, fears of unemployment, and wages demands. Some shipfitters' work was being given to shipwrights, while the shipwrights were at the same time being squeezed out of riveting and certain kinds of machine work by skilled laborers (now 25 per cent of the workforce and the largest single group). Shipwrights and skilled laborers actually worked side by side but for very different wages, even though both were equally proficient and performed the same amount of work.[54]

The skilled laborers, performing work which elsewhere was performed by skilled mechanics, were demanding trade status as well as a pay rise. These men were paid eleven different rates, from 18s. to 32s. per week depending on length of experience and the kind of work; one-third, however, were paid the lowest, and a mere handful the highest, rate. Indeed, more than half probably earned less that 21s. 8d., the amount Seebohm Rowntree in 1899 reckoned to be the minimum needed to provide a man, his wife and three children with "only the bare necessaries of physical efficiency".[55]

It was the unskilled laborers (16 per cent of the total) who were literally paid starvation wages (15s.-19s. per week) and who for this reason excited the greatest sympathy. The Admiralty's attitude to these wretches was (as it had ever been) that, despite the wages, more offered themselves than were needed, so there was no need to pay more. Forwood, a Tory Democrat, took a more benign and enlightened view, however:

> We should not too strictly apply economic doctrine and take advantage of the hard position of the applicants, or the numbers of the unemployed...to pay a wage to an adult labourer that will not afford even to the most frugal a decent house and adequate food.

To the argument that a pay rise would cost too much he replied that "in the end it will mean true economy, as fewer men, well fed and in comfort, may be expected to perform more work".[56]

The Admiralty justified paying less than private yards on the grounds of continuity of employment. This, because of the small Establishment (26 per cent of dockyardmen) relative to normal manning requirements, even hired men now enjoyed, whereas workers in private shipyards were seldom in work more than nine or ten months a year (even in good times). Hudson Kearley, the Liberal Member for Devonport, rebutted this argument in the House of Commons; employment was only continuous, he said,

> [because the government's] normal shipbuilding and maintenance programme is always far in excess of the capabilities of our Dockyards...therefore, continuity of employment, such as it was, was a mere matter of expediency, and nothing for which the government was entitled to fine their workpeople.[57]

Much was made of the countervailing value of incidental benefits like the established men's pensions, the hired men's gratuity upon retirement, and half-pay while recovering from a work-related sickness or injury. Forwood, however, doubted that these benefits brought wages up to the minimum obtaining in private yards and argued that the government was taking unfair advantage of the men's natural desire for security.[58]

Pensions, in fact, were "a serious grievance". Less than half of the established men ever received a pension; 36 per cent died and 18 per cent left the service before reaching the age of entitlement (sixty years). Moreover, the worker alone was entitled to payment; his widow and children got nothing. A man forfeited his gratuity when he was transferred from the hired to the established list, yet time spent on the hired list (sometimes many years) did not count towards a pension.[59]

The Admiralty did little to redress the dockyardmen's grievances, except for a pay rise, totalling £70,000, which satisfied no one. The men's resentment was consequently at fever pitch, and the number of petitions rose to 252, in 1892. Meanwhile, a new grievance had been created in the shape of wage classification.[60]

Wage classification, whereby a man was paid a progressively higher wage according to his proficiency and diligence, was seen as being advantageous and was accepted without complaint by most of the dockyardmen, that is by the new metal-working trades, the skilled laborers and certain others. The shipwrights, however, objected as bitterly now as when the Admiralty imposed it in 1833. Forwood favored classification but even so recommended that the shipwrights should be exempt, for, he said, the purpose of improving wages was "to improve the feeling of the men in regard to the service, and by increased satisfaction to encourage the men to extra exertion", whereas classification had the opposite effect.[61]

The main issues raised between 1890 and 1892 were still unresolved in 1893. Meanwhile, the Trades Union Congress in 1892 called upon the government to pay union wages and legislate an eight-hours day; there was broad agreement at the general election that summer that the government should set an example to private industry; and the Conservatives lost five dockyard-borough seats. Early in 1893, Sir John Gorst, another Tory Democrat, moved in the House of Commons that:

> no person should in Her Majesty's Naval establishments be engaged at wages insufficient for a proper maintenance, and that the conditions of labour, hours, wages, insurance against accidents, provisions for old age, &c., should be such as to afford an example to private employers throughout the country.

The fair wages resolution was accepted by the Liberal Government and unanimously voted by the House.[62]

The Admiralty appointed a committee, chaired by Rear Admiral Charles Fane, the superintendent of Portsmouth yard, to deal with the men's grievance, and most of its recommendations were subsequently accepted. There was a further pay rise of £30,000 (grudgingly approved by the Treasury), and the shipwrights were exempted from classification. It was expected that the unskilled laborers would be awarded a flat rate of 19s. per week, but the Committee, invoking economic doctrine and arguing that a rise of such a magnitude might "lead to considerable disturbance of the wages of other classes of workmen, and add materially to the cost of shipbuilding and repairing", only gave the unskilled laborers an entry wage of 17s., rising to 19s. after two years. This, however, was to flout the fair wages resolution, and the Admiralty was shortly obliged to establish the flat rate anyway. Yet a number of prominent grievances were either not redressed at all or only dealt with partially, and these remained sore spots for years to come. The men were not given a say in setting piece-rates, as was the practice in private yards, an issue which continued to be raised in the petitions year after year. The skilled laborers were not classified as mechanics. Hired time still did not count towards an established man's pension, although he no longer forfeited his gratuity, which henceforth was to be paid to his family if he died before reaching pensionable age. The rule against paying any part of a worker's pension to his family was not relaxed, however.[63]

Adoption of the eight-hours day was another significant outcome of the events of 1890-93, even though the shipfitters and smiths alone mentioned it in their petitions, and the Fane Committee had recommended against it. The shorter working day had as yet been adopted by very few employers, but the Trades Union Congress was pushing for it, and several Liberals who opposed eight-hours legislation either lost their seats or were returned to Parliament with reduced majorities at the 1892 general election. Lord Spencer, the First Lord, and Henry Campbell-Bannerman, the War Secretary, thoroughly canvassed the subject and in the end were won over by the explanation given them by William Mather, a Liberal Member of Parliament, of the highly satisfactory experience of his engineering firm, Mather and Platt, in instituting the eight-hours day. Thus the shorter working day was introduced at the Royal Arsenal, Woolwich in February, and in the dockyards in July, 1894. The effect on British industry was slight, however, and many years passed before the eight-hours day became an issue in private shipyards.[64]

A long quiet now settled over the dockyards and was not broken until 1912-13. The prosperity that followed the shipbuilding slump of 1892-95 was paralleled by a massive expansion of the dockyard workforce to 32,000, and this necessitated new concessions. An established man's time on the hired list was finally counted towards his pension, and some wages were increased.[65]

Organization and management were tightened up after 1885, thanks to Hamilton and the Committees on Dockyard Expenditure. Many, however, of the new arrangements were defective. The Controller was restored to the Board of Admiralty and the Director of Dockyards was made directly responsible to him, thus establishing direct and unambiguous lines of responsibility. Yet little was done to give the chief constructor and chief engineer greater latitude in taking decisions, while the appointment of the civil assistants proved to be an unsuccessful compromise between civilian general management and naval superintendence which in the long run weakened individual responsibility. The most striking developments after 1885 — the reduction of building times and costs — were the result of a more settled technical environment, bigger concentrations of labor, provision of more detailed designs and specifications, more expeditious systems for obtaining materiel, and greatly improved information. At the same time, the problem of work discipline at the turn of the century was as intractable as ever, efforts to deal with it notwithstanding, and industrial relations, although placid on the surface, remained treacherous.

Chapter 9

A New Century and New Ideas, 1900-1914

The new century brought new ideas and far-reaching changes in organization and management, carried out by three outstanding First Lords: the Earl of Selborne (1900-05), Earl Cawdor (March-December 1905) and Winston Churchill (1911-15).[1]

Selborne thought that decentralization was the key to efficiency and economy and proceeded to sweep away much of what remained of eighteenth-century management. The first steps were to devolve greater responsibility to the dockyard superintendents, curtail correspondence and returns, and impress "on our principal officers that what we value is the assistance of their brains and of their eyes, not their signatures." Selborne only imperfectly grasped the dimensions of the problem, however, when early in 1901 he appointed a committee[2] to report on ways of relieving the Controller's department of the enormous burden of unnecessary paperwork that had survived Hamilton's attack and had since grown (and diverted attention from the business of building and repairing ships). The Committee on the Controller's Department reported that even though a body like itself was capable of effecting "many desirable reforms" which otherwise resistance to radical changes might have stifled, the "tendency to multiply the methods of doing work" would inevitably persist unless "the general situation [was continuously] reviewed and unnecessary work dispensed with." Thus was the Committee continued as a standing committee until 1907.[3]

A whirlwind of management changes, the thrust of which was to divide work more rationally and free officials from extraneous or irrelevant functions that interfered with their proper responsibilities, swept through the Controller's department in 1901-02.[4]   The superintendents were authorized to approve estimates of repairs up to the amount of £500 (rather than only £100, the amount established in 1886).   Selborne hoped that this bold measure would lead to repairs and refits being performed less routinely and more economically but knew that success depended on the superintendents' backbone and commonsense.   The superintendents and the inspector of yard accounts were no longer required to obtain the Board of Admiralty's approval for certain routine questions (mostly concerning personnel).   The naval ordinance branch was split off from the Controller's department; however, the Director of Naval Ordinance continued to act under the Controller's direction and in conjunction with the Director of Naval Construction and the Engineer-in-Chief in respect of armaments, gun mountings, and so forth.   The reduction in the number of papers the Controller had to read and sign, made possible by decentralization, was simply enormous — 5000 per annum in the case of ship repairs and refits alone.   At the same time, clerical work was much reduced by abolishing useless returns (e.g. the monthly and weekly returns of the distribution of workmen) and modifying others.   A stop was put to the practice whereby the various branches of the Controller's department routinely acquired duplicate copies of each other's correspondence and to the unauthorized practice (still not suppressed) of certain of the professional officers compiling their own expense accounts. Suppressed also was the practice of routinely forwarding all papers sent in the first instance to the Engineer-in-Chief to the Director of Naval Construction also, whether he was concerned or not, a practice which had in most cases delayed action on the matters in question for at least two days.   Indeed, the Director of Naval Construction was relieved of all administrative work not associated with ship design, a reform only made possible by Sir William White's retirement in January 1902.[5]

Management was leaner at the London offices after 1902, but the economies Selborne hoped for in local management did not materialize. The Navy Estimates having "reached the stupendous sum of nearly 35 millions," he warned in February 1903, were "very near their possible maximum" (they had reached nearly £53 million by the time the country plunged into the Great War).   The dockyard workforce grew from 19,000 to 33,700 (77 per cent) between 1895 and 1905.   Meanwhile, Arthur Balfour, the Prime Minister, in March 1904 warned the House of Commons that "there was extreme danger in the growth and increase of direct employment by the State. He did not like the system." It was at this time also that the extent of shirking in the yards was officially reported to have reached scandalous proportions. Selborne, satisfied that Hamilton's reforms had not gone far enough, in 1904 responded to this new crisis in the affairs of the dockyards by appointing a high-powered committee of inquiry.[6]

The Naval Establishments Enquiry Committee brought together men the outlook of many of whom was in some degree shaped by the example of private industry — Admiral Sir John Fisher (the chairman, whom Selborne had just brought back to the Admiralty as First Sea Lord and who was chairman also of the Standing Committee on the Controller's Department), Rear Admiral Henry B. Jackson (the newly appointed Controller), Sir Evan MacGregor (Permanent Secretary of the Admiralty), Gordon Miller (the new Accountant General), Alexander Gracie (managing director of the Fairfield Shipbuilding and Engineering Company), and Sir James Williamson (Director of Dockyards). Selborne, however, went out to South Africa as Governor General several months before the Committee reported (July 1905), leaving his successor, Cawdor, to carry out its recommendations. Balfour wanted a businessman as First Lord and could not have chosen better than Cawdor, who came to the Admiralty fresh from a highly successful chairmanship of the Great Western Railway.

The redoubtable Fisher, a human dynamo and larger than life (although small in stature), was pre-eminently qualified to be the Committee's chairman, having been Controller himself (1892-1896), commander of the Mediterranean fleet, Second Sea Lord, and a member of Lord Esher's committee on the War Office (1903-04). He had performed with characteristic zealotry on behalf of various controversial causes and now was in his element shaking the Admiralty and the dockyards out of their easy-going ways and prodding everyone to work like "fiends"[7] and get things done.

Vice Admiral William H. Henderson, the superintendent of Devonport yard and a hot reformer influenced by the new management movement that had sprung up in the United States, was one of those who proferred advice as the Admiralty once more wrestled with the question of how the dockyards might be managed as well as commercial shipyards. Henderson was well-read in the literature about British and American systems of organization, had absorbed it all, and was especially impressed with the decentralized systems of the great railways that sprawled across the vast North American continent. This, he told the

Admiralty in November 1904, was the way the dockyards should go.

Loss through the uneconomic use of labor and materials was a problem with which the Admiralty had been wrestling since Sandwich's time at least. Henderson had learnt from management studies that all but a small part of such loss in manufacturing — "5% on labor and 2½% on material" — was the consequence of "faulty and not up-to-date systems, including local inertia," and therefore preventible. Preventible loss in the dockyards came, he reckoned, to a staggering half million pounds annually, a figure that was sure to get attention.[8]

Henderson described the inertia that afflicted Devonport in language with which the reader has by now become familiar.

> When I came to this Yard I found it a huge bureaucracy, in consequence almost nerveless, real energy, initiative or true responsibility had been obliterated; and the object of everyone was to save their face by compliance with the Regulations, or, in default of this, to shift responsibility by reference to the Admiralty. Over regulation and tradition had converted the Departments into separate camps, each struggling for its own, the efficiency of the Yard as a whole was lost sight of, and the economic loss in consequence was very great.

The underlying problem, he said, was a centralized system of organization that was incapable of providing for "the economical administration of the dockyards"; upon a decentralized system alone was it possible for "modern undertakings with all their complexities...[to be managed] economically and efficiently".[9]

The Fisher Committee was convinced of the "imperative necessity for...reorganization of the system of management of the Royal Dockyards, more especially in regard to the supervision of labour and the coordination of the various departments", to be achieved by bringing local management "into line with that obtaining in the best organized private Shipbuilding and Engineering Establishments."[10] Fisher and Henderson were keen to adopt the relatively simple systems obtaining in these; however, A.M. Carlisle of Harland and Wolff a few months later cautioned another committee that the dockyards could never "be worked on exactly the same lines as a private yard"; private firms had only one yard, whereas the Admiralty had a number of yards and required "a far more elaborate system".[11]

Henderson proposed to delegate to the superintendents "the whole of the work, with the power to do it as they choose, and with the power to devise and carry out the best expedients for doing it, without reference or interference." The Fisher Committee, however, far from wishing to hand over to naval officers responsibilities they were hardly qualified to exercise, proposed to curb free-wheeling superintendents who made unauthorized innovations, as Henderson himself crowed about having done at Devonport and saving the Admiralty (he said) tens of thousands of pounds.[12] It was no doubt with superintendents like Henderson in mind that the Committee noted that:

> entire changes in system may be attempted when a new Superintendent enters office, the effects of which might paralyse continuity of system, and so seriously interfere with the discretion of Heads of Departments that they refrain from taking responsibility.[13]

The Committee, dropping the pretence that the superintendent was responsible for the details of departmental management, declared that real responsibility could be exercised by no one "except the officer immediately controlling and acquainted with the work," and that although the chief constructor and the chief engineer were indeed already held responsible for the end result, they were nonetheless denied the authority necessary "for the proper and economic performance of the work". After 1905, therefore, the chief constructor and chief engineer occupied a position like that of managers in private industry and possessed "full authority" within their departments,

> including the power to enter, discharge, promote or punish men (short of discharging men on the establishment), procure their own yard machinery, and get so far as practicable their own stores direct from the contractors under standing contracts without any intermediaries, and control the stock and storage pertaining to their departments.[14]

Their titles were changed accordingly, to manager, constructive department (except at Sheerness and Pembroke) and manager, engineering department. At the same time, the office of civil technical assistant was abolished; it had not produced the intended results and was thought to be inconsistent with the declared intention of making the department heads managers in the full sense. This audacious decentralization was make possible by subjecting the department heads to more rigid financial control, more detailed allocation of the monies voted for the yards, and tighter cost control over large repairs and refits.

Labor and materials were now broken down by trade, for the first time making it possible, when comparing the expense accounts weekly with the approved estimate, to identify those trades which had incurred costs that were over estimate but offset by savings made by certain other trades, and to target problem areas.[15]

The new dispensation was weakened, however, by the failure to appoint general managers at the principal yards. Robinson and Childers had correctly seen that a civilian general manager and a naval superintendent were both needed, each in his different sphere. The Committee, however, confounding the two spheres, dismissed the possibility of appointing what, to their minds, would have been a civilian superintendent. The power of the superintendent continued nominally to be supreme; he was "in the position of the owner (acting on behalf of the Admiralty)"; all matters of importance were referred to him, and he issued all orders for work. His functions were defined, however, as being mainly "general direction and supervision, leaving the management to Heads of Departments, and holding the latter personally responsible... for the conduct of the business of the Departments throughout."[16] Thus did the Admiralty perpetuate and adapt a largely empty arrangement, while at the same time acknowledging that a naval officer was ill-suited to be a general manager.

The Committee expected that the Director of Dockyards would function as a kind of travelling general manager, as had more or less been the intention since the inception of the office in 1872. To make this possible, he was now relieved of all administrative routine; this was transferred to the newly appointed Superintendent of the Dockyard Branch (instructed by the Director of Dockyards but directly responsible to the Controller). Henceforth, the Director of Dockyards was expected to spend all of his time visiting the yards (accompanied by his engineer assistant) and to satisfy himself that they were well managed. His instructions were:

> [to attend] not only to the general organization and equipment of the Dockyards and to the coordination of the work of the various departments, but to the classification and distribution, and check over, labour, as well as the supply, storage, stock, and transportation of materials for Dockyard use...[and to review incidental and establishment expenditure, and] examine and report to the Controller upon the defects of ships requiring large repair, and Dockyard proposals thereto, as well as upon estimates of cost....

He (and the engineer assistant) visited private shipyards and engineering works also, to keep abreast of "developments and improvements in shipbuilding arrangements, &c., and in use of labour-saving appliances." The Director of Dockyards, however, could not substitute for a resident general manager.[17]

Meanwhile, Cawdor had taken the decision to build a new fleet of super-battleships and similar (but controversial) battle cruisers conceived by Fisher, the prototype of which, the *Dreadnought*, was laid down at Portsmouth in October 1905. Four dreadnoughts (the name applied collectively to both types) were to be laid down every year, beginning in 1906. The sheer expense of the Cawdor program dramatically reinforced the need, expressed earlier by Selborne, to squeeze as much as possible out of a sovereign,[18] and a sweeping rationalization of dockyard work was now undertaken, making it possible to axe thousands of hired men in 1905 and 1906.

The practice of massing and employing continuously on ships under construction quite large numbers of workmen was now carried even further. The numbers put to work on the *Dreadnought* were simply enormous and employed even more steadily than on any previous ship. Moreover, the men worked three hours overtime, thereby adding 35 per cent to the working day, while a premium was paid for certain kinds of work (e.g. boilermakers', machine, and electrical) and (it was said) increased productivity about by 20 per cent. The *Dreadnought*, as a result, was launched in an astonishing four months and completed only eight months later, a feat, however, that was never duplicated. The prediction that building time would by these methods fall by one-third was not fulfilled in the dreadnoughts subsequently built down to 1914, although it did fall by a still very impressive 23 per cent.[19] Rapid building had the further advantage of reducing the number of ships that had to be under construction at one time to keep the workforce fully and economically employed and thus reduced the need for building-slips, docks, basin accommodation, machinery and workmen, while at the same time increasing the ability of the yards to deal with repairs. The age of the all big-ship navy had arrived, and the building of the dreadnoughts was consolidated at Portsmouth and Devonport, the only yards capable of accommodating ships of that size. Chatham was disparaged as being "a decaying port unfit for war purposes", and £4 million in improvements were canceled. Henceforth, Chatham and Pembroke only built submarines, while Sheerness, capable of building nothing bigger than sloops, had been downgraded already in 1903. Meanwhile, Fisher persuaded Cawdor to strike off the effective list 154 obsolete ships requiring frequent and extensive repair, thereby lightening this burden on the yards.[20]

The new order, however, levied a heavy human toll: between the autumn of 1905 and the spring 1906, 8000 hired men were discharged (out of a total of 25,000).[21]

Twenty-five of the forty-two dreadnoughts laid down between 1905 and 1914 were of necessity built in private yards, a much higher proportion than of previous battleships.[22] Was there, however, a cost-advantage in doing so? The dockyards had no capital accounts, and the work they performed was so varied that it was exceedingly difficult to apportion establishment and incidental charges amongst the different items of work; thus was it difficult also to ascertain the true cost of a dockyard-built ship. Nevertheless, the Select Committee on Estimates in 1913 was satisfied, "on the whole," with the evidence of Sir James Marshall, the Director of Dockyards, that it cost no more to build in the royal yards than in private yards. There were at this time eight "battleship" firms (out of more than thirty major builders), a smaller number than just a few years before. Building battleships for the Admiralty was not profitable, however; indeed, several of the eight firms made heavy losses doing so, and only three in 1909 were in what was described at the time as a "flourishing" condition, and then only because they were diversified and made a profit from either the manufacture of armor and guns or general manufacturing. Nevertheless, they made a greater loss from idle plant than from a low tender, while Admiralty contracts were essential if they were to win (profitable) contracts from foreign governments and brought business to the profitable branches of the diversified firms.[23]

The physical efficiency of the dockyards was considerably enhanced also. Devonport between 1896 and 1907 was provided with a large and impressive new fitting-out yard at the north end, at a cost of £6 million. Portsmouth, always the most efficient yard, was not surpassed, however; its equipment was fully modernized beginning in 1906, and the yard itself was extended between 1904 and 1914.[24] Electrification of the yards, however, was the most notable improvement of all. Electric power was more adaptable, more easily transmitted, easier to control and adjust, more efficient, and in the long run more economical than steam power, and by 1900 had been widely adopted by the dockyards and private yards alike. Indeed, a number of private yards building the most expensive ships were fully electrified and even had their own central generating plant. The dockyards, and most private yards, had introduced electricity more or less tentatively and partially, however. The dockyard scheme, executed between 1903 and 1908 at a cost of £1 million, electrified all lighting and machinery (except steam hammers and similar appliances), steam cranes, capstans, and so forth, and hydraulic machinery, and provided points around the docks and basins for lighting the ships and driving small portable machines, while each yard was supplied from its own central generating plant. (It was at this time that the electrical engineering department was created.) Electric lighting at last made possible the same working hours the year round and for the first time provided adequate illumination for night work (necessary in wartime) and for overtime in winter.[25] On the other hand, up-to-date and expeditious clerical aids — the typewriter, shorthand and the telephone — were resisted. The Committee on the Dockyard Writing Staff reported in 1908 that:

> the long-hand draft letter, and the laboriously prepared fair copy, are still the order of the day, where, outside, there is now the shorthand note and the typewriter. The hand-sent written message to some other Department still resists the advances of the telephone. That more modern methods have recently been introduced and with great advantage is, of course, not denied; but there is considerable leeway to be made up in this direction.[26]

Store management was further rationalized. The Admiralty confronted two problems: how to reduce the heavy bureaucratic paperwork associated with "duality of control and responsibility" for stores between the Director of Stores and the Controller, and how to prevent the accumulation of redundant inventories. The solution was to make the principal dockyard officers responsible for their own stores and to separate dockyard stores (4659 articles) and sea stores (3201 articles) physically, except for miscellaneous stores (8140 articles) common to both, while leaving the Director of Stores responsible for the administration of all three categories.[27] The unauthorized subsidiary store depots, which the professional officers had been setting up conveniently near the work ever since the present-use storehouses were unwisely suppressed twenty-five years before, were now officially sanctioned and, indeed, mandated.[28]

Doubt was cast on the efficacy of the stock valuation account, introduced by Forwood in 1888, whereupon the account was abolished, and an improved system of stock valuation was adopted. Use of the stock valuation account as an independent check on the store accounts (which were of quantities) necessitated procedures that were very laborious, time-consuming and costly; thus a set of quantity ledgers, corresponding to the storekeeper's, kept independently by the expense accounts department and thought to be more expeditious and effective, was substituted. A comparison of these ledgers was just another paper check, however; inventory-taking was the only reliable check, but only if it was sufficiently extensive and wholly independent of the store department, criteria which the system in use since the 1870s did not satisfy. The local store department staff took inventory, but only of those articles actually in the storekeeper's charge, and there were, besides, test inventory-takings by the London staff; yet neither was thought to be reliable, since it served only the internal purposes of the store department. The checks undertaken by the Accountant General's department were indeed independent, but they were made by a single clerk, who did the same for victualling, naval ordinance, and works department stores also, and were therefore infrequent and limited to only a fraction of the stores each year; nor was the clerk familiar with the stores themselves. The recommendation, made in 1908, that a clerk from the expense accounts department should be substituted (as being more familiar with store transactions), was long resisted by the store department, on the grounds that the two departments were too closely associated on a daily basis, and only implemented in 1914.[29]

The enormous increase in the number of ships built in the dockyards after 1883 (450 per cent) necessitated increasing the Royal Corps of Naval Constructors in 1902. The increase was less than had been requested, however, and Fisher's dreadnoughts, incorporating rapid improvements in design and engineering, placed on the Admiralty constructive staff a burden which could only be supported by shifting more and more assistant constructors from the dockyards (seven in 1909 alone). The dockyards, as a result, were left shorthanded and obliged to revert partially to the pre-1883 system and designate foremen of the yard to oversee construction.   Sir Philip Watts,[30] the Director of Naval Construction, argued that assistant constructors alone were qualified and therefore that the Corps should be increased accordingly, while James Marshall, the Director of Dockyards,[31] argued that many foremen of the yard were just as qualified. Marshall thought, indeed, that the dockyards could be run as well without a constructive corps and that "a good proportion" of assistant constructors could be obtained by promoting foremen, views which reflected the disappointed aspirations of the foremen, who since 1901 were not eligible for promotion.[32]

The Committee on the Royal Corps, chaired by Lord Inchcape (chairman of the P&O and British Steam Navigation companies) was finally appointed in 1910, after two years of deadlock, and agreed with Watts, strongly supported by both his own and Marshall's predecessors, Sir William White and Sir James Williamson. Another thirty-four assistant constructors were appointed (half of them to oversee new work in the yards), the very number that Watts had requested. The Committee's report spoke of how the problems of naval construction were every year becoming more complex and demanding greater "inventive genius and initiative." The Admiralty constructors were ultimately responsible "for maintaining progress," but the role of the dockyard constructors was vital also.

> [T]he shipbuilder at the Dockyard can do much or little to assist the designer; he will do little if he is regarded as merely the instrument for converting approved drawings and specifications into ships; and he may do much in proportion as it is held to be his duty to suggest improvements in details of the design as the work of building a ship advances to completion. It appears to us that the more highly trained a man is, the more naturally he will take this wider view of his responsibility.

The Committee recommended also the advantages of interchanging Admiralty and dockyard constructive staff, as had been intended when the Royal Corps was created in 1883 but was actually done only to a very limited extent.[33]

The Inchcape Committee was concerned that the provisions made in 1883 for broadening the base of the Royal Corps had been ineffective and that the Corps in consequence was drawn almost exclusively from the narrow class of dockyard apprentices. The most important of the Committee's proposals was that one-third to one-half of the places at Greenwich should be reserved for students who had either graduated with distinction in naval architecture from one of the three universities that awarded such a degree, or had taken an honors degree in the mechanical sciences tripos at Cambridge University, and who would be guaranteed membership of the Corps upon successfully completing the Greenwich course. This plan, like the 1883 provision to broaden the Corps, came to nothing, though. Otherwise, it might have helped to reshape the inbred management culture.[34]

The dockyard schools where the Greenwich students prepared had not been a shining example of technical education. In 1903, however, Sir Alfred Ewing, one of the most eminent British engineers of the time, was brought from Cambridge to be Director of Naval Education. Ewing enormously strengthened both curriculum and staff and made the schools more accessible beyond the first year, so as to provide the yards with better-educated supervisors. The schools now became, indeed, an exemplar of technical education.[35]

Most private shipbuilders, on the other hand, still had no use for educated men. Their construction managers (equivalent to assistant constructors in the dockyards) were chosen for their ability to organize, coordinate and control the work and the workmen; they did not need (and did not have) more than a dash of mathematics or theory (just enough to read a plan), acquired in the drawing office and at Board of Education evening classes. They were constantly supervised by the head of the firm, however, and the work, except for that of the "battleship" firms, was simpler than in the dockyards. In fact, the "battleship" firms, with the example and experience of the dockyards before them, now accepted that their construction managers should have "high mathematical attainments."[36] Such men were hard to find, though, largely through the industry's own neglect.[37]

It was generally acknowledged by the time Winston Churchill became First Lord in October 1911 that decentralization, only six years before, had not gone far enough and that the Controller's business had since increased to such a degree that he needed to be given some additional relief.[38] Thus Churchill appointed yet another Committee on the Controller's Department.[39] It was imperative, he said, in an age when "the march of naval science" was leaving "the designs of every year behind it, obsolescent as soon as projected, and when naval tactics and naval strategy" were continually "modified as a consequence of new inventions and developments in materiel," that the Controller should as far as possible be "relieved of routine and administrative functions" (as the Director of Dockyards had been already) and be free to bestow his "undivided attention" and "advise the Board [of Admiralty] upon the supreme subject in his charge" — making certain that the right types of ships were built to carry out the Admiralty's war policy, and that they were completed on time. This could only be achieved by transferring the "vast mass" of civil business — the dockyards, finance, and contracts and purchasing, about much of which the professional experience of an admiral afforded no particular knowledge — to a permanent additional civil lord responsible for furnishing "the Third and Fourth Sea Lords with all that they may require in order to build, arm, equip, and supply the Fleet." The proposal to substitute civilian for naval control over the dockyards was perhaps the most startling of Churchill's ideas.[40]

The report of the Committee on the Controller's Department (July 1912) was faithful to Churchill's ideas. The means of carrying them out, however, were frankly admitted to have been "extremely difficult" to devise.

> On the one hand, it is necessary that the work should be coordinated in all its branches so as to give efficient results, while on the other the work of each branch has become so complex as to require the undivided attention of a specialist. Each department requires to be given full responsibility for its own work, whilst its special functions must be kept in proper relation to the functions of other specialist branches. We have made it our first object to hold this paramount requisite steadily in view, as without it the effective administration of the Board of Admiralty over a most vital part of its wider responsibilities would be seriously impaired.[41]

The Controller's department was abolished, and its various branches themselves became departments, each under a superintending lord responsible for general administration and coordination within a convoluted organizational structure. The Director of Naval Construction, Engineer-in-Chief, Superintending Electrical Engineer, Director of Naval Ordinance, and Director of Naval Equipment (a new position) were all superintended by the Third Sea Lord (without the additional title of Controller), while the Director of Dockyards, Superintendent of Construction Accounts and Contract Work, and Inspector of Yard Accounts were under the new Additional Civil Lord, and the Director of Stores was under the Fourth Sea Lord. The design and construction of ships and alterations and repairs were the province of the Third Sea Lord; the general organization and management of the dockyards, all business connected with the building and repair of ships and their machinery in the dockyards and private yards, and contracts for materiel and stores, were the province of the Additional Civil Lord.[42] The department heads, however, were responsible to two superintending lords — to the Third Sea Lord for technical matters and to the Additional Civil Lord for non-technical matters. For example, the Director of Dockyards was responsible to the Third Sea Lord for building ships "in accordance with specifications" and for duly executing repairs and alterations, but to the Additional Civil Lord for "work and organization," while the Director of Naval Equipment was involved also. Recent building delays, in both the dockyards and private yards, were said to show "that the work must be continuously watched by high naval authority, in respect not only of design, but of progress." Thus the Director of Naval Equipment (a rear admiral) was appointed to follow the progress of naval construction generally and, in association with the Director of Dockyards and Superintendent of Contract Work, to advise the Third Sea Lord accordingly, to prepare annually "a program for additions, alterations, and repairs of ships...[in consultation with] the Director of Dockyards as to the cost and method of carrying out the work in the dockyards," and to confer with the yard superintendents "on all questions of detail affecting the equipment and fitting of...Ships".[43]

Contemporaries who tried to fathom the new organizational system were understandably perplexed. Indeed, Churchill himself had to admit that it was "somewhat anomalous," although only "at first sight". It was, he said, "inevitable", although just why is not obvious.

> [It] presents no difficulties in practice and is well understood by all concerned. The departments are in fact the foundation which unites the different spheres of the Third Sea Lord, the Director of Naval Equipment, and the Additional Civil Lord, and by their common science prevent the risks of technical discordance.[44]

Churchill's reorganization, whatever its shortcomings, nonetheless struck another blow for decentralization. The Director of Dockyards henceforth issued orders on his own responsibility and approved estimates for repairs, alterations and additions, while the director of each department was responsible for its finances.[45]

Churchill went outside the Admiralty organization to select the Additional Civil Lord and chose Sir Francis Hopwood, the Permanent Undersecretary of State for the Colonies, a civil servant of wide and varied experience who had served on a number of royal commissions. It was, however, Hopwood's years at the Board of Trade, as secretary of the railways department (1893-1901) and permanent secretary (1901-07), that chiefly provided him with the credentials Churchill wanted.

The extraordinary organizational system that Churchill thrust upon the Admiralty did not survive the 1914-18 War. The sea lords resented creeping civilian influence, of which the dockyards were the most egregious example. The First Sea Lord, Admiral Sir Henry Jackson, writing on behalf of all the sea lords in November 1915, complained that much of the naval work he did to help the First Sea Lord when he was Controller (1905-08), he now had to do himself as First Sea Lord, "owing to the absence of any Naval officer of experience on the Board being connected with the Dockyards." The "refit question" was in "a distinctly unsatisfactory condition"; there were shortages of materials for repairs, shortages of labor; new construction was "swamping repairs" yet was itself falling behind — all problems that "a far seeing Naval Controller would have watched for and provided against."[46] Hopwood may not have been the man for the job, at least not in wartime, yet most of the fault lay elsewhere. If anyone was to blame for new construction swamping repairs, surely it was Fisher, Jackson's predecessor as First Sea Lord (1914-15),[47] who over formidable opposition had pretty well gotten his own way and "spent much of his energy on the task of constructing a vast armada of ships of every type."[48] Fisher was forced out of office (along with Churchill) in May 1915, but the greater problem remained. This was a war like none before, requiring the total mobilization of vast industrial and technological resources of enormous complexity; failure to do so until after 1916 naturally took a toll on the dockyards and the whole war effort.

The naval war machine was weak, and Lloyd George acted quickly and decisively to strengthen it when he became Prime Minister. Hopwood retired [49] and was succeeded by Eric Geddes, with the title, however, of Controller and with sweeping powers to organize, coordinate and control the country's shipbuilding resources, mercantile as well as naval.[50] It was a job for which Geddes, a railway engineer who had demonstrated organizational skills of a high order in the United States, India, at home, and most recently as Director-General of Military Railways, was admirably suited. He served only a few months, though, before becoming First Lord (July 1917) and was succeeded by Sir Alan Anderson, the head of the Orient Steamship line. The Controllership of 1917 was a wartime expedient which did not validate, even though it continued, the organizational system of 1912. On the contrary, Geddes in November 1918 reunited the offices of Third Sea Lord and Controller.

Renewed parliamentary criticism of the way the government treated the manual workers in its employ, a campaign by the trade unions to organize the dockyards and establish collective bargaining, and rising discontent among the dockyardmen themselves obliged the Admiralty to pay attention to industrial relations.[51] The Boilermakers' Society in 1902 ended its long boycott and voted to organize the yards, and no wonder; the union's rate of growth was slowing down, and the dockyards accounted for one-quarter of the kingdom's shipyard workers. The underpaid and perennially dissatisfied skilled laborers were chiefly targeted; it was argued, by the union and in the House of Commons, that since they did the same work they should have the same trade status and receive the same pay as platers, riveter, caulkers and holders-up (Boilermakers all). The redundancies of 1905-06 made a mockery of the Admiralty's contention that low wages were offset by security of employment, created wide sympathy for the victims, drew attention anew to the issue of wages, and made the hired men more receptive to the unions. Thus the Liberal government in 1906 pledged to "pay the trade union rate of wages paid for similar work" in the dockyard districts and raised wages all round. The skilled laborers, however, were still not classified as mechanics, although those who were discharged received certificates attesting to their skills. Nonetheless, the Boilermakers had enough support to force the Admiralty to let union observers attend the interviews which followed the dockyardmen's annual petitions. The shipbuilding industry shortly fell into one of its worst depressions, however, and the union leaders, moreover, were soon occupied with negotiations for a national agreement with the Shipbuilding Employers' Federation.[52]

The shape of labor troubles was much more ominous between 1911 and 1913. The national agreement broke down; the now booming shipbuilding industry was wracked by strikes and lockouts, and one pay rise followed another in rapid succession. The militancy spread to the dockyards, where the stepped-up naval program and rapid expansion of the workforce to 43,000 gave the men a strong hand. Workers' committees were organized and public meetings were held in support of large pay rises. Even more ominous, the engineering unions at several yards threatened not to work overtime and even to strike; indeed, a number of engineers did refuse overtime, thereby delaying major refits. It seemed as if industrial action might spread, and at the Admiralty there was nervous talk about "the balance of Naval power and...national safety". The rises conceded in October 1913 were inevitable but not generous, lifting wages no higher than the lowest level to which they were likely to fall in private yards when hard times returned.[53]

The Admiralty thought that a desire for steady employment was the chief inducement for men to join a union and that unionism might therefore be neutralized by increasing the number of established men. The Treasury was skeptical, however, saying that it knew of no instance "in which the absence of labour difficulties among the hired class can be shown to have been directly attributable to the effect of an Establishment", and pointing out that for twenty years the Establishment had rarely exceeded 25 per cent of dockyardmen, a proportion too small, in its opinion, to "have much effect in the long run on the large amount of hired labour on which the Dockyards must at all times rely." In fact, the Treasury wanted to axe the Establishment altogether (pensions were unfunded and cost £100,000 per year). The argument over the Establishment went on from 1906 to 1910, when a modest increase, from 6000 to 6500, was sanctioned. The workforce now grew very rapidly, however, and the proportion of established men fell to around 16 per cent. Then came the strike scare of 1913. The Admiralty proposed, and the Treasury reluctantly accepted, an increase to 10,000 (25 per cent), to be reached in two stages by 1915. It would probably have been better, however, if the Establishment had been phased out. For the existence of a large privileged class not subject to the same disciplinary standards as the great majority of workers can only have acted as a brake on disciplinary action generally.[54]

Industrial relations remained unsettled despite the dockyardmen's loyalty. The men were concerned to improve their standard of living, but the Treasury held pay rises to the bare minimum. The real wages of skilled and unskilled laborers (few of whom were established) rose markedly between 1890 and 1913; they did so, however, from oppressively low levels, and the Admiralty was adamant in not conceding trade classification to the skilled laborers. The real wages of the shipwrights, whose versatility made them so valuable, rose hardly at all; the shipwrights were mostly established, however, and disinclined to jeopardize their jobs.[55] Still, an Establishment was not really necessary to ensure loyalty; few shipwrights and skilled laborers could have found comparable employment elsewhere, and most hired men were anyway secure in their jobs. Most men, indeed, when faced with the hard choice, chose security over wages. It was a choice, however, which most would have preferred not to make. They were well able, however, to enforce their own ideas about the just proportion between wages and work, and had, moreover, been doing so for over two hundred years.

The Admiralty in the first decade of the twentieth century built aggressively on the recommendations made by the Committees on Dockyard Expenditure and largely carried out by Hamilton. The object was to decentralize management to the greatest practicable extent. The chief constructors and chief engineers at last became truly responsible managers and were given the title of manager. Managers were relieved of duties that interfered with their main responsibilities, and bureaucratic paperwork was reduced, enabling them, as Selborne hoped, to serve with their brains and eyes rather than with their signatures. The failure to appoint civilian general managers was a serious omission; however, the Director of Dockyards was better able than ever before to oversee and invigorate local management. The quality of information was further improved, cost controls were tightened up, and store management was made more rational. Churchill's contribution was to carry decentralization a step further. His reorganization of the Controller's department, however, was a loser. Conspicuous in all of this was the influence, in one way or another, of the management experience of the railways, the only other organizations that compared with the dockyards in size and complexity.

Another impressive reduction in building time was achieved by reequipping the two major yards (and electrifying all the yards), severely rationalizing production, leaving only Portsmouth and Devonport to build the new leviathans, using overtime, piecework and premiums, and by massing even more men on the construction site. The quality of the workforce may have deteriorated, however, as employment soared to 43,000. The Admiralty on the eve of the Great War was confronted, moreover, by thousands of sullen and dissatisfied dockyardmen, bought off by niggardly pay rises and the hope of making it to an enlarged Establishment.

Chapter 10

Conclusion

Observers of the Royal Dockyards in the eighteenth and nineteenth centuries were struck by the want of order and method in the way the business was conducted, the pervasive slackness, indifference and indiscipline, the low output per worker and the apparent high cost of production. The Admiralty, far from being unconcerned, already in the late eighteenth century was trying to put the management of the yards on a better footing, nor did its efforts ever cease; yet nothing that was done seemed to make any difference. Only very slowly did it dawn that the objects of reform were factors of a bad organizational system, poor information, and the mismanagement of human capital, and that it was these problems which had to be solved first. Until the late nineteenth century, however, all attempts to get the fundamentals straight were defeated by faulty analysis, blinkered conservatism, and bureaucratic inertia.

Samuel Bentham was the first to see that the management problems of the dockyards were systemic. The principles of sound management enunciated by Bentham — individual responsibility, clear and unambiguous lines of authority, full and accurate information — were perhaps beyond the knowledge and ability of his day to put into practice. In any case, the Commission of Naval Revision endorsed the organizational system handed down from the seventeenth century as having stood the test of time and only made some minor adjustments, leaving Bentham to complain about the want of radical reform.

The Royal Commission on the Dockyards in 1861 analyzed the reasons why the yards were ill-managed in terms similar to Bentham's and said that they were like big factories and should be treated as such. Great courage and boldness would have been needed, however, to impose radical reform on an intensely conservative, and now intensely defensive, organization. Sir Spencer Robinson soon demonstrated, in a remarkable analysis, why all reforms up to that time had failed and argued that none would succeed unless a radically different organizational system was adopted. It was essentially Robinson's highly controversial, even though eminently sensible (and modern), ideas that H.C.E. Childers carried into practice when he was First Lord from 1868 to 1871. In doing so, however, Childers botched some parts, and all was subsequently undone. It was left to the Committee on Dockyard Expenditure in 1885 to bring the dockyards to the watershed. Organization and management over the next quarter-century were propelled towards a modern mode on the same principles that Bentham had laid down.

The organizational system, so highly centralized that all authority and (in practice, even if not in principle) responsibility were vested in the top managers residing in London, was the single greatest barrier to good management. It survived, however, almost until the first World War. Well was it call "management by correspondence". The Committee of Dockyard Revision observed in 1848 that insistence upon the literal observance of standing orders and instructions stifled responsibility and encouraged slackness; the Committee on Dockyard Expenditure came to the same conclusion. Only in 1905, however, were the chief constructor and the chief engineer given latitude, similar to that of managers in private industry, to plan and execute the work and similar control over their own stores and departmental finance.

Operations were hampered by vesting the general management of the yard in its principal officers collectively instead of appointing a general manager. Relations between the construction and store departments were often rocky, while the construction and marine engineering departments often worked at cross purposes and, indeed, might not even communicate. Bentham saw the need for a general manager and proposed empowering the superintendent (or commissioner, as he then was), a naval officer. This was just what the Admiralty repeatedly tried, but failed, to do for the next one hundred years. Robinson and Childers, like Bentham, believed that a dockyard could not be managed well without a general manager but, unlike Bentham, were convinced that he must be a civilian. Childers got it wrong, though, and the experiment he launched in 1870 failed. Not until 1905 did the Admiralty come round to Robinson's and Childers's view that a naval officer, necessary though his presence was, was nonetheless unsuited to manage the highly technical business of a dockyard. The possibility of appointing a civilian general manager was dismissed, however, as being incompatible with the managerial authority that was now conferred on the chief constructor and chief engineer. Thus the problem of coordination, even though somewhat attenuated by this time, continued to be an impediment to efficiency. There was as yet no general manager in 1914; nevertheless, such an appointment eventually was inevitable.

It was not possible that business as extensive, technical, and complex as that of the Navy Office could be conducted in its entirety, and down to the least detail, by a committee constituted, as the Navy Board was, on the principle of collective authority and responsibility and not fall into confusion. The division of the business in 1796 among three specialized subcommittees and the full Board recognized that a problem existed but was not a solution. The subcommittees were abolished and the Navy Board was streamlined in 1829; the Board, however, only survived for three more years.

The rationale for abolishing the Navy Board (although not the Principal Officers, except the Comptroller) was that the dockyards would be managed better if the Admiralty exercised direct responsibility and control. What happened, though, was that management, far from improving, actually deteriorated. The new structure was organized on the same principles as the old; now, however, the Principal Officers were each superintended by one of the junior lords of the Admiralty, and all orders and instructions had to be approved either by the full Board or one of its members. Moreover, the Principal Officers might communicate with each other only circuitously through their superintending lords. Departmental isolation, delay, misunderstanding and muddle were inevitable, and the Surveyor of the Navy often found himself in the untenable position of one who was held responsible without being given corresponding authority. It was a system that was bound to break down in the event of a long naval war, like those of the eighteenth century, and in fact nearly did break down in the Crimean War. Childers's solution in 1868 (following Robinson) — putting the Controller (formerly Surveyor) of the Navy on the Board of Admiralty, thereby establishing individual responsibility, and subordinating the store department to him — had the merit of being logical and of working also. It was too radical, however, to survive Childers's premature retirement, and the old organizational system was revived, in a modified form. Nevertheless, the Controller was sensibly restored to the Board in 1886.

The process of differentiating functions began with creating, in 1848, a separate branch for design, so as to enable the Surveyor to give more attention to the dockyards. At the same time, the hitherto independent marine engineering department was made a branch also. The appointment after 1864 of inspectors of dockyard work provided the Admiralty with better control and the yard constructors with badly needed guidance. The dockyard branch, created in 1872, was an overdue arrangement which for a long time, however, did not produce all of the benefits that had been anticipated. The Director of Dockyards, to be fully effective, needed to visit the yards frequently and give advice on difficult management and production problems, as, indeed, was originally intended but not fully realized until the Director was relieved of administrative duties in 1905. The last step in the process of decentralization was taken in 1912, when the Director was empowered to issue orders on his own responsibility and to approve estimates for repairs, alterations and additions.

It would surely be thought extraordinary today if the chief executive officer of a big company never visited its factories, yet that was the case with the dockyards for most of the eighteenth century. It was, significantly, the Board of Admiralty, rather than the Navy Board, that initiated the practice of dockyard visitations. The depressingly similar tale of the minutes suggests, however, that whatever beneficial results the visitations of the late eighteenth and early nineteenth century may have had, they were mostly short-lived. Nonetheless, the Admiralty obtained in this manner much useful information enabling it to institute various, even if not always efficacious, reforms. The appointment of the Director of Dockyards and the inspectors of dockyard work so diminished the usefulness of the visitations, however, that they were very nearly discontinued after 1900.

The number of accounts and records which the dockyards were required to compile and transmit to London in the eighteenth century might suggest that the usefulness of such information to management was well understood, yet, paradoxically, none of this voluminous information was either very accurate or subjected to serious scrutiny by the persons for whom it was compiled. The appointment of the dockyard accountants in 1856 reflected a new perception of the role of accountancy, and was followed by the discovery that the expense accounts were wholly unreliable. The errors were long thought to be clerical, and it was only in 1878 that they were demonstrated to be the consequence of faulty accounting practices; accounting practice did not improve sufficiently until after 1885, however, for financial control to be established.

The expense accounts were expanded and overhauled in the 1860s but, now being under the Accountant-General of the Navy and his agents the yard accountants, were not very accessible either to the chief constructors or the Controller. Moreover, they were chiefly compiled with a view to satisfying the interest of the House of Commons and the Treasury and thus were of no practical use for following current expenditure. Nor was the allocation of indirect expenses to ships, now attempted for the first time, very successful.

The expense accounts in 1887 were transferred to a newly created branch of the Controller's department, and a test audit was instituted. It was now possible (for the first time) to follow current expenditure from (reasonably) accurate accounts and thus to control it also. The effect on the yard officers was soon evident in the way they vied with each other to control costs and in better estimating. After 1905, the yard officers were subjected to closer financial control, monies were allocated in greater detail, the expense accounts were refined, and tighter cost control was established. It was, indeed, owing to developments in accountancy that the Admiralty thought it safe to entrust the chief constructor and chief engineer with control over their own finances. There were as yet, however, no capital accounts in 1914, and difficulties were still being encountered in apportioning indirect expenses to ships.

Store management continued to be very weak despite the continuous survey instituted in 1849. Sweeping changes occurred in the 1870s, however. The continuous survey, hitherto indifferently observed, was entrusted to the yard accountants; variable, self-adjusting store establishments were substituted for the old (and long since obsolete) permanent establishments; and a new system of accounts provided more accurate information about current inventories. It was now possible to substitute a single annual demand for the immemorial quarterly demand, and occasional demands, which had had long since been routine, were no longer permitted. Inventory control and the disposition of stores were vastly improved; unannounced, random surveys of the balances in the storekeeper's ledgers were instituted in 1886 and subsequently refined; a useful stock valuation account, first attempted in 1864, remained elusive, however.

People are an organization's most valuable asset but an asset that was badly managed in the eighteenth century and even worse managed in the 1820s and 1830s. Inadequately educated naval constructors, uneducated supervisors, jobbery and favoritism in making promotions, and low wages all took their toll.

The School of Naval Architecture in its several incarnations played an important role in professionalizing the constructive corps. The talents of many of the graduates were misapplied, however, and it was only with the creation of the Royal Corps of Naval Constructors in 1883 that the transition from craft to fully professional status and mentality was completed, with a perceptible effect on morale.   Meanwhile, the dockyard schools from the 1840s were preparing the brightest apprentices for the School of Naval Architecture and the next tier for sub-managerial positions. The system of recruitment and training, however, produced an unhealthy degree of inbreeding, especially since the apprentices (and thus supervisors and managers) were mostly the sons of dockyardmen. The Admiralty tried to attract new blood to the School of Naval Architecture, but none was forthcoming.

Merit, ascertained largely by written examination, was adopted in 1847 as the criterion for promotion to supervisory positions, thus putting an end (eventually) to jobbery and favoritism. The apprentice schools and the merit system were expected to provide the yards with intelligent and competent first-line supervisors; first-line supervision continued, nevertheless, to be slack, nor do successive reorganizations appear to have helped much.

It was to compensate for low wages, which otherwise might have made it difficult to attract and retain good skilled workmen, that the practice of employing men for life was adopted by the yard officers and condoned by the Navy Board. Superannuation, introduced in 1764, but for the first fifty years applied selectively, was the Admiralty's response to the inefficiencies created by employing workmen for life. It was not until 1855, though, that a retirement age was mandated, and only when the age was lowered to sixty years in 1860 that the yards were entirely relieved of their burden of old men. Meanwhile, the Admiralty in 1833 adopted a bipartite division of the workforce into permanent, pensionable men and temporary, non-pensionable men (around 75 per cent of the workforce by the 1890s) who, however, might hope eventually to be placed on the former list. This arrangement was seen as an instrument of subordination but nevertheless created much dissatisfaction in both groups.

Slack work-discipline and industrial unrest, problems still at the beginning of the twentieth century, were above all factors of low wages, nor were any of the various stratagems employed to increase the intensity of labor very successful; some, indeed, only spurred resentment. The men felt degraded by the pilfering to which they were driven by low wages in the eighteenth century and by the treatment to which they were compelled to submit in the 1820s and 1830s. The shipwrights felt betrayed when they adapted to metal shipbuilding but were denied the higher wages they thought had been promised. The policy, adopted in the late nineteenth century, of substituting cheaper for more expensive labor whenever possible provoked new grievances. The differential between wages in the dockyards and private yards was much less by the twentieth century; however, it was the Admiralty's policy to keep wages at the lowest level to which they were likely to fall elsewhere in a trade slump. If there was slackness, it was no wonder.

Employment was virtually hereditary. The shared experiences of generations of workmen, living in tight-knit communities and belonging to extensive "clans", based on consanguinity, marriage, or some other connection, helped to shape a dockyard culture. The background of so many supervisors was the same; they either identified with, or, living in those same communities themselves, were compelled to identify with, labor rather than management. These were facts which no amount of systemic or structural change could have modified very easily. The officers generally preferred to work cautiously within the parameters of the dockyard culture rather than to assert themselves and upset the delicate balance between work and authority.

How, then, are we to explain Lord George Hamilton's boast that the dockyardmen were the fastest and most efficient shipbuilders in the world and Sir James Marshall's assurances that building costs were competitive with those in private yards? A more rational organization of labor (e.g. continuous working), more detailed drawings and specifications, better accounts and records, better financial control, the institution of cost controls, a fully professional constructive corps, well-educated foremen of the yards, closer Admiralty contact with the yards, the war on bureaucratic paperwork, and, perhaps most important of all, decentralization — all played a part. Most occurred after 1885; none would have been possible, however, except for earlier groundwork, mostly after 1848. It was, significantly, the centralized system of "management by correspondence" that survived the longest.

Of course, we shall never know if the royal yards were either as efficient or as economical in 1914 as was claimed, since overheads were inevitably greater than in commercial yards (owing to the unique variety of the work), and allocation of indirect expenses was problematical. Nor do we know how much it actually cost the "battleship" yards to build for the Admiralty; we do know, however, that they made a loss doing so. As for repairs and maintenance, there was broad agreement that these were better and more cheaply performed in the dockyards, although the basis of comparison was rather slender. What is certain, however, is that the royal yards in 1914 were better managed than in 1885 and much better managed than in the eighteenth century. That was an impressive achievement considering the very heavy burden, inherited from the distant past and sanctioned by long usage, that had to be thrown off. Today a less favorable verdict has been rendered, and the dockyards have come under private management.

Notes

Chapter 1: Introduction.

1. John Brewer, *The Sinews of Power. War, Money and the English State, 1688-1783* (Cambridge, Mass., 1990), p. 34.

2. Westcott Abel, *The Shipwright's Trade* (Jamaica, N.Y., 1962), pp. 105-06; H.C. (1910) lxi. 594-95; H.C. (1914) liii. 454-455; Oscar Parkes, *British Battleships, "Warrior" to " Vanguard"* (London, 1956), pp.16-24, 562; Phyllis Deane and W.A. Cole, *British Economic Growth 1688-1959. Trends and Structure* (Cambridge, 1962), p. 330.

3. James M. Haas, "The Royal Dockyards: The Earliest Visitations and Reform 1749-1778," *Hist. Journ.*, XIII (1970), pp. 210-14; ADM 7/593, pp. 525-311; ADM 1/5614, 3 Aug. 1852; 58 & 59 Vict., c. 35 (1895); 3 Edw. 7, c. 22 (1903).

4. C.S.L. Davies, "The Administration of the Royal Navy under Henry VIII: the Origins of the Navy Board," *Eng. Hist. Rev.*, LXXX (1965), pp.268-69, 276-77.

5. Sidney Pollard, *The Genesis of Modern Management* (London, 1968), p. 12.

6. Navy Board to Secretary of Admiralty, 21 March 1708/09, R.D. Merriman (ed.), *Queen Anne's Navy. Documents concerning the Administration of the Navy of Queen Anne*, (Navy Records Society, 1961), p. 82; ADM 7/662, fos. 7-8; ADM 106/2200, 11 May, 1771, unfoliated pages at end.

7. ADM 3/79, 11 Aug. 1772; ADM 106/2201, p. 293; *Rept. of the Committee on Contracts for the Building or Repairing of Ships*, pp. 1-3. H.C. (1884), xiv. 1; Charles Oscar Paullin, *Paullin's History of Naval Administration 1775-1911* (Annapolis, Maryland, 1968), pp. 396, 474.

8. *Rept. of the Select Committee on the Navigation Laws. Evidence*, q. 5304. H.C. (1847-48) xx. 498; Leslie Jones, *Shipbuilding in Britain* (Cardiff, 1957), p. 77.

9. D.C. Coleman, "Naval Dockyards under the Later Stuarts," *Econ. Hist. Rev.*, 2, VI (1953), p. 139; *Queen Anne's Navy*, p. 373.

10. Pollard, *Modern Management*, pp. 71-72.

11. A.E. Musson, *The Growth of British Industry* (New York, 1978), p. 199; Jones, p.77 and n.1; S. Pollard, "The Decline of Shipbuilding on the Thames," *Econ. Hist. Rev.*, 2, III (1950), p. 88.

12. ADM 1/5716, *Rept. of the Committee on Navy Estimates* (1859), p. 8.

Chapter 2: The Eighteenth Century.

1. Daniel A. Baugh, *British Naval Administration in the Age of Walpole* (Princeton, NJ, 1965), pp. 293-99; *1st Rept. of the Commission of Naval Revision*, pp. 25-88. H.C. (1805), ii. 25.

2. *3rd Rept., Commission of Naval Revision*, pp. 40-42. H. C. (1806), v. 282; ADM 106/3240, fos. 146-47; Samuel Bentham, *Services Rendered in the Civil Department of the Navy* (London, 1813), pp. 38-39; Baugh, pp. 293-99. The situation of American managers of Japanese factories in the United States is somewhat similar. They are "accustomed to individual responsibility" and find it frustrating not to be "allowed to make decisions or fully use their talents." "Each morning...U. S. managers at Mazda [Flat River, Mich., works] get a 'grocery list' from their Japanese 'advisers' that tells them just what they are supposed to do that day." *Wall Street Jounal,* 27 Nov. 1991, p.1 and 17 April 1990, p. B7.

3. The actual intructions to visit Deptford and Woolwich seem, however, to have been issued only in 1749. ADM 2/215, 10 Nov. 1749.

4. A commissioner for these two yards was appointed for a short time during the War of the Austrian Sucession.

5. *6th Rept. of the Commission on Fees*, p. 11. H. C. (1806), vii. 106; *6th Rept. of the Commission of Naval Enquiry*, pp. 3-4. H. C. (1803-04), iii. 3; ADM 2/265, pp. 472-73; ADM 3/61, 10 Aug. 1749; John Knox Laughton, ed., *Letters and Papers of Charles, Lord Barham* (3 vols., Navy Records Society, 1907-11), II, p. 333; Baugh, pp. 287-93; Roger Beckett Knight, "The Royal Dockyards in England at the time of the War of America Independence," Ph.D. diss., University of London (1972), pp. 51-53.

Baugh is inclined to credit the Commissioner at Portsmouth, Capt. Richard Hughes, with the relatively good state of that yard when the Board of Admiralty visited it in 1749; Portsmouth, however, generally comes off well in later inspections also.

6. *Rept., Commission on the Control and Management of Naval Yards,*

Appendix, no. 53 (Memorandum by Rear Admiral Sir Spencer Robinson, p. 589). H. C. (1861), xxvi. 589; ADM 1/6104, 16 Jan. 1869 (8 Feb. 1869, fo. 1).

7. ADM 106/2201-2509 (1772-82) *passim.*

8. Davies, pp. 268-69, 276-77.

9. The Treasurer of the Navy was originally a member of the Navy Board but soon dropped out. His office became a sinecure and was executed by the Paymaster until the Paymaster's department was merged in the new department of the Paymaster-General in 1835.

10. Francis H. Miller, *The Origin and Constitution of the Admiralty and Navy Boards* (London, 1884), pp. 6-7; Baugh, pp. 32-33, 35-38.

11. Baugh, pp. 40, 48, 89-92; ADM 3/61, 26 June and 7 July 1749.

12. John Ehrman, *The Navy in the War of William III, 1689-1697* (Cambridge, 1953), pp. 195-97; Baugh, p. 84; ADM 3/80, fo. 201.

13. Baugh 72-84; N. A. M. Rodger, *The Admiralty* (Lavenham, 1979), pp. 64-65.

14. *Barham Papers,* II, pp. 236-41.

15. ADM 106/2509, S. O. 289 (20 March 1784); ADM1/3462, 11 June 1822; *Barham Papers, II, pp. 239-40;* Nathaniel Barnaby, *Naval Development of the Century* (London, 1905), p. 435.

Paul Webb speaks of "casual methods and a relaxed attitude to paper work," yet suggests that "further study remains to be done in our understanding of the Navy's accounting procedures." (Paul Webb, "Construction, repair and maintenance of the battle fleet of the Royal Navy, 1793-1815," in Jeremy Black and Philip Woodfine, eds., *The British Navy and the Use of Naval Power in the Eighteenth Century* [Atlantic Highlands, NJ, 1989], p. 218.) It is unlikely, however, that the endeavor, if it is ever undertaken, will lead anywhere other than down a blind alley.

16. ADM 106/2182, 24 May 1746.

17. *2nd Rept. of the Commission of Naval Revision,* pp. 3-8. H. C. (1806), v. 248; ADM 1/3462, 8 Dec. 1821.

18. ADM 106/2507-2508 (Standing Orders) *passim*; ADM 2/215, 10 Nov. 1749; ADM 3/61, 16 June-12 Aug. 1749; *6th Rept., Naval Enquiry*, pp. 5-6; *1st Rept. of the Commission of Naval Revision*, pp. 8-9. H.C. (1805), ii. 8.

19. ADM 3/36. 26 June 1749; CL, Shelburne MSS., Vol. 151, nos. 55-61.

20. IND 9310, nos. 3, 7, 8; ADM 2/237, 22 April 1767.

21. Pollard, *Modern Management*, pp. 71-76.

22. ADM 2/215, 10 Nov. 1749; ADM 2/237, 23 June 1767; ADM 3/61, 10 Aug. 1749; ADM 3/62, 2 June 1752; ADM 3/75, 11 June 1767; ADM 106/2186, 2 March 1752; ADM 106/2196, 7 Sept. 1764.

23. Piers Mackesy, *The War for America, 1775-1783* (London, 1964), pp. 19-20; James M. Haas, "The Pursuit of Political Success in Eighteenth-Century England: Sandwich, 1740-71," *Bull. Inst. Hist. Res.,* XLIII (1970), pp. 56-77.

24. ADM 7/662, fos. 73-74.

25. See James M. Haas, "The Royal Dockyards: The earliest Visitations and Reform 1749-1778," *Hist. Journ.*, XIII (1970), pp. 191-25.

26. ADM 2/215, 10 Nov. 1749.

27. See the minutes of the visitation. ADM 3/61, 26 June - 12 Aug. 1749.

28. ADM 106/2507, S.O. 399 (13 March 1750).

29. Egmont was First Lord from 1763 to 1765.

30. ADM 2/234, 18 June 1764.

31. *Barham Papers*, II, p. 212.

32. First Lord 1766-71. Hawke, the victor over the French at Quiberon Bay in 1759, was one of the most distinguished admirals of the century.

33. ADM 2/237, 22 April 1767; ADM 3/75, 11 June 1767; ADM 106/2198, 5 June 1767; ADM 106/2199, 17 Nov. 1769.

34. ADM 7/659-662.

35. CL, Shelburne MSS, Vol. 146, no. 105.

36. ADM 7/662, fo. 72.

37. *Ibid.*, fo. 68.

38. CL, Shelburne MSS, Vol. 146, no. 105.

39. *Ibid.*, Vol. 146, no. 174.

40. ADM 7/662, fo. 72.

41. ADM 106/3222 (consisting of the Navy Board's rough notes made on the spot); ADM 2/269, 12 Dec. 1792.

42. This is clear from the minutes of the visitation of 1813-14 made just after Sandwich's long lost accounts of his own visitations had been discovered. ADM 7/593, inscription on fly-leaf and pp. 41, 62.

43. Melville visited the yards in 1813-14, 1816 and every year thereafter.

44. ADM 1/7735, 23 Sept. 1904; ADM 1/7814, 21 Dec. 1905 ("Official Procedures and Rules," 1 April 1905), pp. 56-57.

45. Mackesy, pp.163-65.

46. Julian S. Corbett and H. W. Richmond, eds., *Private Papers of George, second Earl Spencer* (2 vols., Navy Records Society, 1913-14), I, p. 6.

47. *Barham Papers*, II, pp. 208, 212, 300-01, 423-24, 426, 428-29.

48. *Ibid.*, II, pp. 209-10, 216, 237-38, 338-39, 423-24; CL, Shelburne MSS., Vol. 151, no. 40, fos. 13-14; ADM 2/265, p. 308.

49. *Barham Papers*, II, pp. 198-208, 209-3, 236-41, 344; *6th Rept., Naval Enquiry*, pp. 6, 134; ADM 106/2509, S.O. 37, 270, 327.

50. Ministry of Defence, Naval Historical Library, N. Macleod, Extracts of Portsmouth Dockyard Records, 1 Aug. 1729; ADM 1/381, Mathews, 17 Sept. 1742; ADM 3/62, 2 June 1752; ADM 7/662, fos. 71-72; ADM 1/3462, 8 Dec. 1821; ADM 106/2198, 17 June 1767; ADM 106/2508. S.O. 918, 1125; Bentham, *Services*, p. 39; NMM, POR F/6, 22 Feb. 1740.

51. ADM 1/3477, 27 Aug. 1832; ADM 1/2283, Chatham, 8 Aug. 1832; ADM 3/61, 6 July 1749; ADM 106/2198, 17 June 1767; ADM 7/662, fo. 60; *6th Rept., Naval Enquiry*, pp. 10-11; *Repts. of the Committees on Dockyard*

Expenditure, p. 17. H.C. (1886), xiii. 155.

52. Pollard, *Modern Management,* p. 173.

53. R.J.B. Knight, "Sandwich, Middleton and Dockyard Appointments," *Mariner's Mirror,* 57 (1971), pp. 179-84.

54. ADM 1/5977, 6 Dec. 1866 (Controller's memo., 1 Dec. 1866); *3rd Rept., Naval Revision, Appendix,* no. 95; *Rept. of the Select Committee on Dockyard Appointments,* p. xiii. H.C. (1852-53), xxv. 25; *Barham Papers,* p. 27.

55. *Rept. of the Committee on Dockyard Economy,* p. 21. H.C. (1859 [2], xviii. 21; ADM 116/330 (Report by A. Forwood, 4 Dec. 1890, p. 5); Hansard, 4, ix, cols. 1163, 1168 (6 March 1893); ADM 7/662, fos. 5-6.

56. The effect on subordination seems not to have been perceived until the 1840s, however. H.C. (1847),xxxvi, 142-43.

57. Sandwich's appointment books have survived but shed no light. NMM, SAN/1-3, 5-6. See, however, M.J. Williams, "The Naval Administration of the Fourth Earl of Sandwich," D. Phil. diss., Oxford University (1962), pp. 277-82.

58. Knight, "Dockyard Appointments," pp. 179-84; *Barham Papers,* II, pp. 11-30; ADM 106/1809 (Chatham, 5 July 1801).

59. *6th Rept., Commission on Fees, Appendix,* no. 144.

60. *3rd Rept., Naval Revision,* pp. 180-81, *Appendix,* nos. 43-86; *Barham Papers,* II, p. 11; ADM 106/2208, p. 430.

61. Baugh, p. 304.

62. Pollard, *Modern Management,* pp. 127-47.

63. James Pritchard, *Louis XIV's Navy 1748-1762. A Study of Organization and Administration* (Kingston and Montreal, 1987), ch. 6 and 7.

64. *3rd Rept., Naval Revision,* p. 193, *Appendix,* no. 45.

65. *3rd rept., Naval Revision,* p.203.

66. Of the firm of Fletcher and Fearnall, Union Dock, Limehouse. The firm built a number of ships, both sail and steam, for the Admiralty between 1831 and 1856. Philip Banbury, *Shipbuilders of the Thames and Medway* (Newton Abbot, 1971), pp. 175-76.

67. ADM 1/3462 (1822); ADM 1/3476, 24 March 1832. The Surveyor's estimate has been omitted because his figures were based on a new and untried task scheme he was promoting. He would have employed 192 shipwrights and paid £5760 in wages.

68. ADM 1/5977, 22 Nov. 1866 (26 Nov. 1866).

69. Samuel Bentham, *Answers to the Comptroller's Objections on the Subject of His Majesty's Dockyards* (London, 1800), p. 20; *6th Rept., Naval Enquiry,* p.13.

70. Neil McKendrick has inferred from practices developed by Sir Josiah Wedgwood that cost accounting may have been wider spread in the late eighteenth century than has been thought; however, the scale and complexity of naval construction would not have lent themselves to such simple methods. Moreover, external stimuli, in the shape of severe economic pressure in

Wedgewood's case, were altogether lacking in the case of the dockyards. Neil McKendrick, "Josiah Wedgewood and Cost Accounting in the Industrial Revolution," *Econ. Hist. Rev.*, XXII (1970), pp. 45-67. See also S. Paul Garner, *Evolution of Cost Accounting to 1925* (University, Alabama, 1954), pp. 29-30, 69.

71. Henry Roseveare, *The Evolution of the British Treasury* (London, 1969), pp. 88-92; J.E.D. Binney, *British Public Finance and Administration 1774-92* (Oxford, 1958), pp. 139-50; Baugh, pp. 464-70; A.C. Littleton, *Accounting Evolution to 1900* (2nd edn., New York, 1966), p. 320; David Solomons, "The Historical Development of Costing," in David Solomons, ed., *Studies in Cost Analysis* (2nd edn., Homewood, Ill., 1968), pp. 4, 8.

The Navy Debt "consisted of 'Navy Bills' — quite simply the unpaid receipts of the Navy for goods, services and victuals." Roseveare, p. 93.

72. ADM 3/44, 3 Feb., 18 Apr., 8 Sept. 1740; ADM 106/3377, 19 June 1740; ADM 1/3495, Admiralty memo., Feb. 1838; *3rd Rept., Naval Revision*, pp. 194-95; *Rept., Naval Yards, Evidence*, qq. 9932-9933, *Appendix*, Robinson's Memo., p. 596; *Barham Papers*, III, p. 20; Bentham, *Answers to Comptroller*, p. 18; Barnaby, p. 435; Knight, "Royal Dockyards," pp. 330, 353; Baugh, pp. 338-39.

73. Knight, "Royal Dockyards", pp. 330, 336; Baugh, pp. 337-38; ADM 7/593, pp. 127-28; ADM 1/6060, 2 March 1868.

The time needed to get to Spithead, etc. was considerable, the men might be seasick, and bad weather might hinder work for days.

74. "Of the 55 projects in excess of £0.5 million started and completed in the first two years of the Dockyards' commercial management, 39 were completed late." *Report by the Comptroller and Auditor General, Ministry of Defence: Fleet Maintenance* (London, 1990), p. 14.

75. Knight, "Royal Dockyards," pp. 336, 342-43, 353; Baugh, pp. 334-35; Roger Morriss, *The Royal Dockyards during the Revolutionary and Napoleonic Wars* (Leicester, 1983), pp. 18-22; ADM 7/593 (1813), p. 137.

76. *3rd Rept., Naval Revision, Appendix*, nos. 91-93; ADM 7/662, fol. 5; ADM 106/2203, 26 July 1775, p. 160; ADM 106/2508, S.O. 673 (23 March 1775).

77. ADM 116/84, 9 June 1896.

78. Baugh, pp. 314-5.

79. *3rd Rept., Naval Revision, Appendix*, nos. 90, 92; I.J. Prothero, *Artisans and Politics in Early Nineteenth-Century London* (Folkestone, 1979), p. 37.

80. *8th Rept. of the Commission of Naval Revision*, p. 69 (the *8th Report* [1807] was not printed; all references are to the copy in Ministry of Defence, Naval Historical Library); ADM 7/662, fos. 37-38, 65; ADM 7/663, pp. 158-59; ADM 2/125, 10 Nov. 1749; ADM 3/61, 10 Aug. 1749.

81. ADM 3/61, 10 Aug. 1749; ADM 106/2184, 28 May 1748; ADM 106/2196, 30 Oct. 1764; W[illiam]. S[hrubsole]., *A Plea in Favour of the*

Shipwrights in the Royal Dockyards (Rochester, 1770), pp. 20, 29; *Barham Papers*, II, pp. 199-200, III, p. 18.

82. ADM 106/2196, 3 Oct. 1764, 21 Nov. 1764; ADM 106/2201, 21 Aug. 1771; ADM 2/269, 12 Dec. 1792; ADM 2/301, 19 Feb. 1803; ADM 2/261, 18 Feb. 1785; ADM 106/2514, No. 34 (2 March 1803); ADM 7/593, pp. 68, 252-53, 597.

83. The hours changed more than once in the nineteenth century. Only the electrification of the yards in 1908 made possible a uniform working day year-round.

84. *6th Rept., Naval Enquiry*, pp. 9-10; ADM 1/3462 (Fearnall to Martin, 15 Dec. 1821; "Earnings in Kings Yards and Merchant Yards and Hours of Work").

85. ADM 1/5991, *Rept. of the Committee of Revision — Dockyard* (1848), fols. 44-45; *Rept., Naval Yards, Appendix*, no. 29 (General Orders and Board Minutes, 25 Jan. 1849, p. 575; *Rept., Dockyard Economy*, p. 15; ADM 1/5892, 31 Dec. 1864; ADM 1/6588, *Dockyard Regulations* (1882), pp. 141-42.

86. Wages were paid quarterly and until 1772 were usually as much as a year in arrears, thereby obliging the men to borrow at the high rate of 5 per cent. James M. Haas, "The Introduction of Task Work into the Royal Dockyards, 1775," *Journ. Br. Studies*, VIII (1969), p. 61.

87. BL, King's MSS. 44, fos. 5-32 *passim*.

88. ADM 3/62, 2 June 1752; *6th Rept., Naval Enquiry, Appendix*, nos. 108-19; A.L. Bowley and George H. Wood, "The Statistics of Wages in the United Kingdom during the Nineteenth Century (part XIV). Engineering and Shipbuilding," *Royal Statistical Soc. Journ.* (1906), p. 184; Pollard, *Modern Management*, pp. 165-72.

89. ADM 106/3306, Woolwich, 28 Oct. 1743; Shrubsole, p. 13.

90. There is no contemporary evidence; I have extrapolated these figures by assuming that task-work earnings were the same per cent more than the underlying day rate as in the earlier nineteenth century. *Rept., S. C. on the Navigation Laws, Evidence*, qq. 6047, 6133.

91. *11th Rept., Commission on Crown Lands, Commons Journals*, xlvii (1792), p. 372.

92. ADM 7/79, fo. 141; ADM 7/703, 10 May 1765; ADM 7/662, fo. 36; Shrubsole, p. 13; *6th Rept., Naval Enquiry*, pp. 9-10; Bentham, *Answers to Comptroller*, p. 107; NMM, POR F/8, 25 April 1747.

93. ADM 1106/2181; NMM, ADM B/198, 5 Jan. 1779.

94. *Rept., S.C. on Navigation Laws, Evidence*, qq. 6047, 8091-8095, 8334, 8337.

95. I.J. Prothero is of the opinion that the London shipwrights "assumed that older people...earned the right to a decent standard [of living]", since just after the turn of the century, when runaway price-inflation was inflicting severe hardship, they tried unsuccessfuly to persuade the builders to give work at a reduced wage to the graybeards worn out by years of work. (Prothero, p. 37.)

The shipwrights were not so high-minded as Prothero thinks, however; they were responding to pressure and trying to shift the burden from their friendly societies (i.e. from themselves) to the builders. Not only were they unwilling to sacrifice a part of their own wages (as was done in the dockyards), but they themselves were part of the problem, since they excluded the older men from their gangs.

96. *8th Rept., Naval Revision*, p. 69; ADM 7/662, fos. 37-38; ADM 3/67, 12 Dec. 1759.

97. Shrubsole, p. 19.

98. B.R. Mitchell and Phyllis Deane, *Abstract of British Historical Statistics* (Cambridge, 1962), pp. 346-47.

99. Shrubsole, pp. 2-22; ADM 7/703, 10 May 1756; ADM 106/2201, 14 Feb. 1772.

100. Sandwich introduced task work for joiners, bricklayers and housecarpenters in 1772-74, and certain kinds of unskilled laborers' work had been performed by the task since 1758. ADM 106/2508, S.O. 458 (3 April 1758), 604 (11 June 1772), 610 (2 Sept. 1772), 619 (23 Dec. 1772), 667 (31 Aug. 1774), 669 (16 Sept. 1774).

101. ADM 106/2186, 2 March 1752; ADM 7/662, fos. 2, 4, 7-8, 9, 54, 60-61; G.R. Barnes and J.H. Owen, eds., *The Private Papers of John, Earl of Sandwich*, (4 vols., Navy Record Society, 1932-38), IV, pp. 287, 310-11.

The task scheme and regulations were printed in *3rd Rept., Naval Revision, Appendix*, nos. 107-109.

102. ADM 7/662, fos. 4-6; ADM 106/2508, S.O. 673 (23 March 1775).

103. The original plan was to lay down a single ship in each yard, but there were no vacant slips. ADM 3/81, fo. 10; ADM 106/2508, S.O. 673 (23 March 1775).

104. ADM 7/662, fos. 23-29, 34-35, 46, 51-53, 63-65; ADM 106/2203, 17 June 1775, 6 and 26 July 1775; ADM 3/81, 17 and 18 July 1775; ADM 2/243, 4 Aug. 1778; ADM 106/2202, pp. 203-04, 209-11; *Sandwich Papers*, IV, p. 288; *Gentlemen's Magazine*, XLV (1775), pp. 403-05.

For a fuller account of task work and the strike, see Haas, "Task Work," pp. 44-68.

105. ADM 7/662, fos. 23-25, 34-37, 47, 63; ADM 2/243, 4 Aug. 1775; ADM 106/2203, 17 June 1775, 6 July 1775, 5 Sept. 1775.

106. ADM 106/2508, S.O. 918 (3 Nov. 1779); ADM 106/2203, 5 Sept. 1775; *Sandwich Papers*, IV, p. 288.

107. I have chosen to use the term "piecework" rather than "job work", the official term for over a century, in order to avoid confusion later on when the task and job schemes become identical. Piecework/job work, like task work, was adapted from commercial yards.

108. *Sandwich Papers*, IV, p. 409; ADM 106/2205, 7 March 1778; ADM 106/2509, S.O.76 (12 April 1783) and 236 (3 Dec. 1783).

109. *6th Rept., Naval Enquiry*, pp. 20-23; *8th Rept., Naval Revision*, p. 79;

ADM 106/2509, S.O. 76 (2 April 1783), 100 (15 May 1783), 236 (3 Dec. 1783), 249 (18 Dec. 1783), 672 (28 Dec. 1791); ADM 106/2211, 22 Nov. 1783; *11th Rept., Crown Lands*, p. 372; Mitchell and Deane, p. 347.

110. Hansard, 4, ix, cols. 1163, 1168 (6 March 1893).

111. Shipwrights in commercial yards enjoyed the same privilege.

112. ADM 3/61, 6 July 1749; ADM 106/2198, 17 June 1767; *11th Rept., Crown Lands*, pp. 279, 372; Samuel Bentham, *Naval Papers and Documents* (London, 1827), pp. 277-80, 282-83; ADM 106/2211, 11 Sept. 1783; ADM 106/3234, 22 Aug. 1820; IND 12375, 41.20; H.C. (1860), xiii. 323.

113. ADM 106/2211, pp. 183-84, 260-80, 298.

Chips had been commuted to wages in the case of all other workers as long ago as 1649. *Commons Journals*, VI, pp. 381-82 (13 March 1649).

114. ADM 1/6273, pt. 1, 9 June 1873.

115. ADM 3/61, 29 June, 5 and 6 July 1749; *1st Rept., Naval Revision*, p. 3; *Barham Papers*, II, p. 239; Knight, "Royal Dockyards," pp. 220-22, 227.

116. Even in 1908, no less than fourteen steps were followed in receiving stores. ADM 1/7990, 14 Jan 1908 (*Rept. of the Accounts Subcommittee of the Naval Establishments Committee*, pp. 20-21).

117. It was for this reason that the clerks upon entry were willing to pay the seemingly astronomical premium of 150-250 guineas to the chief clerk of the department and each time they were promoted (by seniority) to pay the clerks whom they succeeded. The discovery of a serious case of embezzlement at Portsmouth led Sandwich in 1773 to abolish the sale of clerkships in the store department and require the store clerks to give bond (as the storekeeper himself did) and to increase their salaries. ADM 2/242, 5 May 1773; ADM 3/80, 31 March, 5 May and 12 Aug. 1773.

118. *6th Rept., Commission on Fees*, pp. 231-39, 282-84, 305, 314, *Appendix*, nos. 32, 65-67, 100, 133, 185-86, 235-37; *1st Rept., Naval Revision*, pp. 44-45; Knight, "Royal Dockyards," pp. 241, 243.

119. *1st Rept., Naval Revision*, pp. 44-45; Bentham, *Answers to Comptroller*, pp. 311-32; Knight, "Royal Dockyards," pp. 241, 243.

120. ADM 1/5591, 14 Dec. 1848, *Rept., Dockyard Revision*, fo. 12; *5th Rept. of the Commision on Fees*, p. 310, *Appendix*, nos. 16-17, 161; H.C. (1806), vii. 310; *1st Rept., Naval Revision*, p. 45; Knight, "Royal Dockyards," pp. 241-43.

121. ADM 106/2508, S.O. 764 (30 Oct. 1778), 771 (30 Nov. 1778), 821 (1 March 1779), 864 (22 June 1779), 916 (28 Oct. 1779), 992 (21 June 1780); ADM 106/3404, 7 Dec. 1778; ADM 106/3240, 26 Oct. 1810, fos. 65-66; ADM 106/3462, 8 Dec. 1821; ADM 1/5591, 14 Dec. 1848, *Rept., Dockyard Revision*, fo. 6; *6th Rept., Commission on Fees, Appendix*, nos. 4, 170, 222.

122. The plan was drawn up by Palliser and presented to Sandwich as soon as the latter became First Lord. *Sandwich Papers*, IV, p. 279.

123. ADM 7/659, fos. 54, 85; ADM 7/660, fos. 12-13, 48-49, 84-85; ADM

7/661, fo. 75; ADM 7/662, fos. 18, 19, 31, 42, 50, 57, 70, 74-75; ADM
106/2186, 22 Jan. 1753; ADM 2/244, 2 Aug. 1776; ADM 2/261, 18 Feb. 1785;
ADM 49/124; BL, Add. MSS. 38217, fo. 275 (Sandwich to Jenkinson, 21 Jan.
1782); Knight, "Royal Dockyards," pp. 216-20.

124. ADM 7/593, pp. 104-05, 292-93; ADM 1/5591, *Rept., Dockyard
Revision*, fo. 38; *1st Rept., Naval Revision*, pp. 8-10, 611; *Rept., Naval Yards*,
p. 8.

125. ADM 7/593, pp. 194-95, 465, 484; ADM 7/661, fos. 2, 26, 42-43, 62;
ADM 7/662, fos. 11, 31, 75; ADM 106/2508, S.O. 567 (4 April 1771); ADM
106/3240, 26 Nov. 1810, fos. 92-93; *11th Rept., Crown Lands*, pp. 278-79;
Shrubsole, pp. 10-12; Bentham, *Services*, p. 1.

126. ADM 106/2196, 21 Nov. 1764; ADM 1/5167, 21 June 1765; NMM,
ADM B/165, 7 and 21 July 1760; Richard Middleton, "Naval Administration in
the age of Pitt and Anson, 1755-1763," in Black and Woodfine, eds., p. 112.

127. ADM 106/296, 2 Nov. 1764, pp. 330, 348-55, and unfoliated pages
following; ADM 7/659, fos. 25, 60; ADM 7/660, fos. 34-36, 54; ADM 7/661,
fo. 2; ADM 7/662, fos. 78-79; ADM 2/261, 18 Feb. 1785, pp. 94-96; ADM
2/262, 4 May 1786; BL King's MSS. 44, fos. 3, 8, 12, 20, 27, 32; NMM,
ADM B/183, 21 Aug. 1770.

For a fuller account, see Haas, "Royal Dockyards: Visitations and Reform,"
pp. 210-14.

128. ADM 7/659, fos. 24-25, 60-61; ADM 7/660, fos. 5-7, 28-30, 31, 34-46,
45, 65, 81-82; ADM 7/661, fos. 27, 39-42, 46-48; ADM 7/662, fos. 30, 39, 78;
Barham Papers, III, p. 32.

129. ADM 7/593, pp. 194-95, 372, 465, 484, 523; *Repts., Dockyard
Expenditure* [1886], *Evidence*, q. 1157.

130. Pollard and Robertson, p. 109.

Chapter 3: Tinkering with the System, 1793-1815.

1. Binney, pp. 15, 251.
2. *Barham Papers*, II, pp. 302-04.
3. Middleton served briefly under Spencer as First Naval Lord.
4. ADM 2/265, 17 May 1796.
5. See R.A. Morriss, "Samuel Bentham and the Management of the Royal
Dockyards, 1796-1807," *Bull. Inst. Hist. Res.*, LIII (Nov. 1981), pp. 226-40.
6. See the biography by his widow, M.S. Bentham, *The Life of Brigadier-
General Sir Samuel Bentham* (London, 1862); Samuel Bentham, *Services
rendered in the Civil Department of the Navy* (London, 1813), pp. iv-v.

Bentham was never given a British knighthood but was permitted, as a
courtesy, to use his Russian title.

7. Bentham virtually designed his own office. He did not get everything he
wanted (his ideas were rather grandiose); nevertheless, he had a staff of seven,

including an architect and engineer, a machinist and a chemist.

8. *Ibid.*, pp. 131-35; Samuel Bentham, *Naval Papers and Documents* (London, 1827), pp. 4-10, 21-36, 56-142, 265-76; ADM 1/5188, 23 March 1796; *Repts. from the Committees of the House of Commons*, xiii, *17th Rept. of the Select Committee on Finance*, p. 328, *31st Rept. of the Select Committee on Finance*, p. 494; Bentham, *Letters*, p. 36; Morriss, *Admiralty*, pp. 226-27.

Middleton appears to have played a crucial role in creating the office of Inspector-General. He thought the Surveyor and master shipwrights unfit to plan and supervise civil construction and was instrumental in the appointment in 1778 of a civil engineer whose office Lord Howe later abolished as being unnecessary. Middleton then convinced the Commission on Fees to recommend the appointment of a civil architect but later (1795), responding to pressure for the appointment of an inspector general of civil and naval construction such as the French had, proposed creating for this purpose a board intermediate between the Admiralty and the Navy Board. *Barham Papers*, II, pp. 246-47; *6th Rept., Commission on Fees*, pp. 314-315.

9. Bentham, *Services*, pp. 36-41; Bentham, *Letters*, pp. 34-36, 129-30.

10. There can be no doubt that the idea originated with Samuel rather than Jeremy. See Morriss, *Dockyards*, p. 239, n. 59.

11. B.B. Schaffer, "The Idea of the Ministerial Department: Bentham, Mill and Bagehot," *Austral. Journ. Politics and Hist.*, III (1957-58), pp. 60-61.

12. Abbott was Deputy-Speaker, and later Speaker, of the House of Commons.

13. 17th Rept., S.C. on Finance [1797]; p. 328, *31st Rept., S.C. on Finance* [1797], *passim*.

14. *Barham Papers*, III, pp. 32, 56.

15. Nominally to the master shipwright of the yard after 1801 and to the Admiralty itself after 1832.

16. The dockyard officers, however, were entitled to chips until 1879. ADM 1/6496, 6 Sept. 1879.

17. ADM 1/5194, 21 May 1801; ADM 106/2513, 29 June 1801; H.C. (1801), v. 67-85.

18. This was not the first time serious irregularities had been reported at Plymouth. See *Barham Papers*, II, pp. 333-35.

19. Joseph Tucker, one of the assistant master shipwrights and brother of St. Vincent's personal secretary, Benjamin Tucker. He was later Surveyor of the Navy (1813-31).

20. David Bonham Smith, ed., *Letters and Papers of Admiral of the Fleet the Earl St. Vincent whilst First Lord of the Admiralty 1801-1804*, (2 vols., Navy Records Society, 1922-27), II, pp. 181-83, 193-95, 203, 536, 540; ADM 7/593, pp. 311-12.

It has been suggested that Middleton may have wanted to abolish the Navy Board. Admiral John Markham, one of the junior lords, certainly did, and so also did Charles Grey when he was First Lord in 1806. Clements Markham, ed., *Selections from the Correspondence of Admiral J. Markham* (Navy Records Society, 1904), xv; *Barham Papers*, III, p. 384; J.T. Ward, *Sir James Graham* (London, 1967), p. 128.

21. See John R. Breihan, "The Addington Party and the Navy in British Politics 1801-1806," in Craig L. Symonds, ed., *New Aspects of Naval History* (Annapolis, MD, 1981), pp. 163-89.

22. He was subsequently acquitted.

23. *Barham Papers*, III, p. 73.

24. ADM 106/2539, 29 July 1805; M.S. Bentham, p. 236.

25. *4th Rept. of the Commission of Naval Revision*, pp. 12-13. H.C. (1809), vi. 12.

26. He retired to France after the war and spent the rest of his life (he died in 1831) vainly trying to win the recognition he thought he deserved but had been denied.

27. *1st Rept., Naval Revision*, pp. 8-88; *2nd Rept., Naval Revision*, pp. 3-117; Bentham, *Services*, p. 36.

28. ADM 106/3240, fos. 1-2; ADM 7/593. Melville had been First Lord since 1812.

29. *6th Rept. of the Select Committee on Finance*, p. 90, *Appendix*, p. 211; H.C. (1817), iv. 212.

Some of Seppings' duties were similar to those of the recently dismissed Bentham. The three surveyors met as a committee on naval construction.

30. This office was created on the outbreak of war in 1793.

31. The number of commissioners was increased to ten, and certain offices disappeared, viz. Clerk of the Acts (now the Board's secretary), Comptroller of the Store Accounts, and Comptroller of the Victualling Accounts.

32. ADM 1/5188, 8 June 1796; ADM 2/279, 17 Aug. 1796; ADM 106/2511, 25 Aug. 1796; *4th Rept., Naval Revision*, pp. 27-44.

33. St. Vincent in 1803 issued a similar order pursuant to a recommendation of the Commission of Enquiry, but it may not have been executed. ADM 106/2232, 20 Oct. 1803.

34. *6th Rept., Naval Enquiry*, pp. 29, 62, 66; *2nd Rept., Naval Revision*, p. 173; *4th Rept., Naval Revision*, pp. 6, 57-58.

35. *6th Rept., Commission on Fees*, p. 306; Bentham, *Naval Papers*, pp. 252-54; *31st Rept., S.C. on Finance* [1797], pp. 487, 491, 494-95; ADM 1/5194, 21 May 1801; *1st Rept., Naval Revision*, pp. 11-12, 17-18; ADM 1/3462, Sir B. Martin's Observations (C).

36. Bentham, *Answers*, pp. 102-03; *8th Rept., Naval Revision*, pp. 1038, 1046, 1051-53, 1056; Richard Vesey Hamilton, ed., *Letters and Papers of Admiral of the Fleet Sir Thomas Byam Martin* (3 vols., Navy Records Society, 1898-1903), III, p. 388.

37. Bowley and Wood, p. 189.

38. Mitchell and Deane, p. 470; *6th Rept., Naval Enquiry,* p. 109.

39. *Ibid.,* pp. 22-23, 26-27, 30-31, 39, 46, 79-80, 84, 24, *Appendix,* no. 136; Bentham, *Answers,* pp. 79-80, 83-84.

40. *6th Rept., Naval Revision,* pp. 24, 26, 38, 47-48; ADM 106/2513, S.O. 349 (2 Jan. 1802); Bentham, *Answers,* p. 77.

41. Bowley and Wood, p. 189.

42. ADM 106/2227, 1-4 April 1801 *passim*; ADM 106/1916, 31 March-21 April 1801 *passim*; ADM 106/3223 *passim*; R.A. Morriss, "Labour Relations in the Royal Dockyards, 1801-1805," *Mariner's Mirror,* LXII (1976), pp. 337-46.

43. *6th Rept., Naval Enquiry, Appendix,* no. 136; *8th Rept., Naval Revision,* p. 522.

44. ADM 2/306, 18 Oct. 1804; ADM 2/312, 4 Sept. 1807; ADM 2/317, 31 July 1809; ADM 106/2236, 27 March 1805; ADM 106/2237, 7 Aug. 1805; *6th Rept., Naval Enquiry,* pp. 127-28; *3rd Rept., Naval Revision, Appendix,* nos. 71-73.

45. *6th Rept., Naval Enquiry,* pp. 132-33; *3rd Rept., Naval Revision,* p. 204; *8th Rept., Naval Revision,* pp. 24, 57-68; ADM 3/324, 25 March 1812; ADM 7/593, p. 83.

46. ADM 7/593, pp. 249-50.

47. *6th Rept., Naval Enquiry,* pp. 25-26, 129-32, 146; ADM 106/2235, 15 Sept. 1804.

48. *3rd Rept., Naval Revision,* pp. 202, 205, 211; *8th Rept., Naval Revision,* pp. 8-119, *Appendix,* pp. 454-94 *passim*, 528-38;ADM 2/313, 1 Jan. 1808.

49. Used to ascertain tonnage, from which the ship's price was reckoned.

50. This seems to have been the measure used by commercial builders but was probably just as unsatisfactory. See D.J. Roberts, *The Economics of Wages and the Distribution of Income* (London, 1961), p. 15.

51. *3rd Rept., Naval Revision,* pp. 199-205, 211-12, *Appendix,* nos. 16-47, 110.

52. *3rd Rept., Naval Revision,* pp.209-10; *8th Rept., Naval Revision,* pp. 24, 94, 100, 105-06, 108; ADM 2/306, 3 Oct. 1804; ADM 2/318, 11 Oct. 1809; ADM 2/321, 11 Dec. 1810; ADM 2/323, 8 June 1812; ADM 7/593, pp. 74-75, 351, 621-24; ADM 106/2522, S.O. 679 (2 Sept. 1811), 721 (4 Oct. 1811); ADM 106/2523, S.O. 279 (28 May 1812); ADM 106/2525, S.O. 71 (1 Nov. 1814); *6th Rept., S.C. on Finance* [1817], *Evidence,* p. 212; ADM 1/3462 ("Earnings in Kings and Merchant Yards").

53. *3rd Rept., Naval Revision,* pp. 195, 198, 209-11; J.R. Parkinson, *The Economics of Shipbuilding in the United Kingdom* (Cambridge, 1960), p. 177.

54. Bentham, *Answers,* pp. 115-16; *Rept., Naval Yards,* p. 8.

55. It was Joseph Tucker, then master shipwright at Plymouth, who orginally proposed this method. Bentham appears to have favored it too. *3rd Rept., Naval Revision,* pp. 205, 207; Bentham, *Answers,* pp. 89-90.

56. *3rd Rept., Naval Revision,* pp. 206-07; ADM 2/321, pp. 448-49; ADM

1/3462, item S; ADM 1/3480, 22 June 1833; ADM 7/593, pp. 404-05; *Rept., Naval Yards, Appendix*, no. 34.

57. ADM 2/308, 3 Aug. 1805; ADM 2/325, 30 June 1812; ADM 7/593, pp. 72-74, 81, 440, 663; Morriss, *Dockyards*, pp. 206-07.

58. *6th Rept., Commission on Fees*, pp. 307-08; Bentham, *Answers*, p. 100; Bentham, *Services*, pp. 12-14; ADM 1/5194, 21 May 1801; ADM 2/300, 25 Nov. 1802; ADM 106/2513, 7 Dec. 1802; Morriss, *Dockyards*, pp. 112-23; *Rept., Dockyard Economy*, p. 118.

59. Bentham, *Answers*, p. 110; Bentham, *Services*, pp. 112-14; *6th Rept., Naval Enquiry*, pp. 117, 124; *3rd Rept., Naval Revision*, pp. 187-84, 192, *Appendix*, pp. 308-09; ADM 2/324, 4 Jan. 1812.

60. *3rd Rept., Naval Revision*, pp. 180-83, 193.

61. See Pollard, *Modern Management*, pp. 123-47.

62. *3rd Rept., Naval Revision, Appendix*, no. 45.

63. Bentham, with whom the idea of educating the shipwright officers originated, was light years ahead of his contemporaries in the matter of curriculum: he would have added subjects like the management of shipbuilding materiel, bookkeeping and personnel management. Bentham, *Naval Papers*, "Plan for a System of Instruction," end paper following p. 284.

64. *3rd Rept., Naval Revision*, pp. 183-90, 192-93; ADM 2/320, "Plan of Education...for a Superior Class of Shipwrights," following p. 75; ADM 2/321, Nov. 1810, pp. 273-76; ADM 1/5225, 30 Jan. 1816.

65. Bentham, *Naval Papers*, p. 256; Bentham, *Services*, p. 4; ADM 1/5194, 21 May 1801; ADM 106/2513, 29 June 1801; ADM 106/3240, 26 Nov. 1810; ADM 2/299, 2 April 1802, 23 Sept. 1802; *9th Rept. of the Commission of Naval Enquiry*, p. 10. H.C. (1805), ii, 10; *2nd Rept., Naval Revision*, p. 43; Morriss, "Bentham," pp. 231-32; ADM 106/2515, 6 June 1804; ADM 2/322, 30 May 1811; ADM 7/593, pp. 194-95, 345.

66. *6th Rept., Commission on Fees*, p. 310; Bentham, *Naval Papers*, pp. 254-55; ADM 1/5194, 21 May 1801; *9th Rept., Naval Revision*, p. 81; ADM 106/3240, 26 Oct., fos. 65-83; ADM 7/593, pp. 262, 312; Morriss, *Dockyards*, pp. 85-90.

67. Webb, pp. 208, 210-11; Morriss, *Dockyards*, p. 30; ADM 1/6211, 7 March 1871; ADM 7/593, pp. 126-27, 135-36; ADM 2/326, 26 Nov. 1812. Morriss is surely mistaken in assuming that the increase in tonnage built in the dockyards after 1805 reflected rising efficiency rather than a change in policy.

68. ADM 2/326, 26 Nov. 1812; ADM 2/331, 5 Nov. 1814; ADM 106/2525, 22 April 1814.

69. ADM 7/593, pp. 192-227. Rennie found much fault with Portsmouth, especially with the layout, and drew up his own plan of additional improvements which was obliquely critical of Bentham for not having done better.

70. Bentham, *Naval Papers*, pp. 22-34, 217-20, 236-28, 241-43; ADM 7/593, p. 28; Morriss, "Bentham," p. 236; Morriss, *Dockyards*, p. 54.

71. Bentham, *Services*, pp. 75-78; Samuel Betham, *Desiderata in a Naval Arsenal* (London, 1814), pp. 5-43; ADM 7/663, p. 121-25, 139-40; ADM 7/593, pp. 482-92, 522-33, 545-63; ADM 106/2232, 9 July 1803; ADM 2/325, 5 June 1812; *8th Rept. of the Select Committee on Finance*, pp. 53-55, 58. H.C. (1818), iii. 153; Morriss, *Dockyards*, pp. 53-61.

72. ADM 2/305, 23 April 1804; ADM 2/307, 30 March 1805; ADM 106/1883, 23 May 1805; ADM 106/2516, 24 Sept. 1805; ADM 7/593, pp. 109-17; Morriss, *Dockyards*, pp. 51-52; Morriss, "Bentham," pp. 234-35.

73. ADM 2/326, 24 Dec. 1812; ADM 83/54, 12 Feb. 1849; ADM 106/2525, 21 June 1814.

74. Morriss, *Dockyards*, pp. vi-vii, 218, 221-22.

75. Webb, p. 218.

76. *Repts., Dockyard Expenditure* [1886], p. 21.

Chapter 4: Lightening Ship, 1815-1834.

1. Melville resigned when Canning became Prime Minister in April 1827; Wellington, however, persuaded him to return to the Admiralty in September 1828.

2. The Duke of Clarence (the future King William IV) was Lord High Admiralty advised by a council.

3. ADM 1/3462, 8 Dec. 1821.

4. The Commission of Naval Revision had recommended overmanning as the best way to keep discipline, since every man would know that he could be discharged and his loss would not be felt. *3rd Rept., Naval Revision*, p. 214.

5. All who so wished were conveyed free of charge in revenue cutters to the port nearest their town of origin so as not to burden the local parishes.

6. "Agents of foreign employers scoured Britain [in the first half of the nineteenth century] in the hope of enticing skilled men to France, Russia, Germany, America." E.P. Thompson, *The Making of the English Working Class* (New York, 1963), p. 246.

7. ADM 1/5765, 2 April 1896; ADM 7/593, p. 623; ADM 106/2527, 5 Feb. 1816, 6 March 1816; ADM 106/2283, 25 June 1822; ADM 106/2284, 24 July 1822; ADM 106/3235, pp. 7-10; ADM 1/3462, 20 June 1822; H.C. (1831-32), xxvii. 574-75; BL, Add. MSS. 41368, fos. 318-19, 329; ADM 3/223, 29 June 1831; ADM 106/2296, 1 July 1831; ADM 1/3480, 22 June 1833.

8. ADM 106/2527, 6 March 1816, 4 May 1816; ADM 106/3235, pp. 7, 10-12, 20; ADM 1/3462, 11 and 20 June 1822. The time rate for shipwrights was at first 4s. 6d. per (short) day; in 1822, however, a shortened working week of four days was introduced and the time rate was increased to 5s. Piece wages averaged 6s. 6½d. just before check measurement was introduced in 1816.

9. Melville had the Surveyor, Sir Robert Seppings, draw up a new task scheme that could be used for repairing ships also; this scheme proved to be

impracticable, however. ADM 1/3461, 1 March 1830; ADM 1/3475, 22 Nov. 1831; ADM 106/2296, 1 July 1831.

10. An Admiralty committee reported in 1848, when the workforce was a little smaller than in 1814, that 220 measurers would be needed if all work was actually to be measured. Sixty-two measurers were appointed in 1811, a number that subsequently fell along with the workforce. ADM 1/5591, fo. 32; ADM 1/5667, 8 Dec. 1855.

11. ADM 106/3234, 22 Aug. 1820; ADM 1/3475, 22 Nov. 1831; *Rept., Naval Yards, Evidence,* qq. 2802-2811; ADM 1/3462, 8 Dec. 1821; ADM 1/3477, 9 June 1832, 27 Aug. 1832; ADM 1/3480, 22 June 1833.

12. *6th Rept., Naval Enquiry,* pp. 124-25; ADM 1/3477, 27 Aug. 1832; ADM 1/5243, 18 July 1833.

13. See J.M. Haas, "Work and Authority in the Royal Dockyards from the Seventeenth Century to 1870," *Proc. Amer. Phil. Soc.,* 124 (1980), pp. 422-24.

14. *3rd Rept., Naval Revision,* p. 214.

15. ADM 106/3234, 7 Dec. 1820; ADM 106/3235, pp. 32-36; ADM 1/3462, 20 June 1822, pp. 3-4, 9-10; ADM 1/5132 (Woolwich, 22 Aug. 1822; Deptford, 21 Aug. 1822; Chatham, 26 Sept. 1822); ADM 3/206, 18 Dec. 1824; ADM 106/2296, 1 July 1831; ADM 1/3478, 3 Nov. 1832; ADM 83/8, 3 Oct. 1833; ADM 1/3495, 14 July 1837; BL, Add. MSS. 41368, fos. 277-78 (paper by Martin, c. 1830).

16. ADM 106/2295, 22 Jan. 1830.

17. ADM 1/3480, 22 June 1833; ADM 1/5243, 18 July 1833; IND 12148, 41.8; IND 12200, 41.14; *Rept., Naval Yards, Appendix, no. 16* (Rept. on the Dockyards by R.S. Bromley, 13 Feb. 1847, p. 485).

Although there is little contemporary evidence, the effect of classification on morale may be gathered from the shipwrights' reaction when a similar system was introduced around 1890. See J.M. Haas, "Trouble at the Workplace: Industrial Relations in the Royal Dockyards, 1889-1914," *Bull. Inst. Hist. Res.,* LVIII (1985), p. 26.

18. The work in commercial shipyards was not measured.

19. Reduced in size to ten men and apprentices.

20. Portsmouth yard consisted at this time of 118 acres, Chatham of 90 acres.

21. ADM 1/3462, 8 and 24 Dec. 1821, 11 June 1822; ADM 1/5235, 8 Feb. 1822; ADM 3/200, 18 July 1822; ADM 1/3477, 27 Aug. 1832; H.C. (1831-32), xvii. 575; ADM 1/3480, *Dockyard Instructions* (1833), pp. 41-57.

22. The size of each gang was increased from ten to fifteen.

23. ADM 1/3475, 22 Nov. 1831; ADM 1/3477, 27 Aug. 1832; ADM 1/3480, *Dockyard Instructions* (1833), pp. 117-30.

24. A sharp reduction from the thirty years previously required.

25. It is unlikely that any of the men discharged in 1822 were entitled to a pension.

26. ADM 1/5221, 6 Dec. 1814; ADM 106/3235, pp. 43-46; *Rept. from the*

Select Committee on the Navy Estimates, pp. l-li. H.C. (1847-48), xxi (i). 742.

27. ADM 106/3235, pp. 43-46; *3rd Rept. of the Select Committee on Public Income and Expenditure,* p. 17. H.C. (1828), v. 495; ADM 1/3477, 27 Aug. 1832.

28. ADM 106/3231, pp. 37-47; ADM 106/3232, 23 Sept. 1818; ADM 116/2530, 15 Feb. 1891; ADM 1/5235, 8 Feb. 1822; ADM 1/3470, 27 Nov. 1829.

29. ADM 106/2527, 29 Aug. 1816; ADM 106/2528, 21 March 1817, 10 Sept. 1817; ADM 1/3470, 27 Nov. 1829; ADM 1/5591, 14 Dec. 1848, fos. 12-14.

30. ADM 1/3462, 8 and 24 Dec. 1821; ADM 1/3477, 27 Aug. 1832; ADM 106/2527, 29 Aug. 1816.

31. ADM 1/5580, 27 Feb. 1847, p. 10; ADM 1/3470, 2 Nov. 1829; ADM 1/3477, 27 Aug. 1832.

32. Martin insisted on this exception on the grounds that it was often necessary to divert workmen at short notice.

33. ADM 1/3462, 8 Dec. 1821, 24 Dec. 1821, 11 June 1822, "Sir Byam Martin's Observations" (A, D, E, F).

34. ADM 1/5243, 27 June 1832.

35. ADM 1/5591, 14 Dec. 1848, fo. 3.

36. Christopher Lloyd, *Mr. Barrow of the Admiralty. A Life of Sir John Barrow 1764-1848* (London, 1970), pp. 98-100.

37. Hansard, 3, x, cols. 349-50, 356-58, 361-64, 366 (14 Feb. 1832).

38. Rodger, *Admiralty,* p. 105; Oswyn A.R. Murray. "The Admiralty," *Mariner's Mirror,* XXIV (1938), p. 458.

39. ADM 1/3477, 9 June 1832 (General Instructions); ADM 1/3478, 26 Sept. 1832.

40. The move was to houses in Spring Gardens adjacent to Admiralty House.

41. Rodger, *Admiralty,* p. 99.

42. Sir Eric Geddes upon going to the Admiralty Board during the First World War was astonished "at the lack of any method of conducting Board meetings — the vagueness, the lack of preparation, and the absence of minutes, the amateurishness of the whole". Quoted in Arthur J. Marder, *Portrait of an Admiral: Life and Papers of Sir Herbert Richmond* (London, 1952), p. 267.

43. *Rept., Naval Yards,* Robinson's Memo., p. 592.

44. Extended to the Army and Ordnance departments in 1848.

45. *Repts., Dockyard Expenditure* [1886], pp. 5-6.

46. ADM 1/3479, 1 Jan. 1833, J. Briggs, *Report on the French and English Systems of Public Accounts, Chiefly in the Naval Departments.*

47. ADM 1/3480, 22 June 1833, *Dockyard Instructions,* p. 128; ADM 1/5591, 14 Dec. 1848, fols. 41-42.

48. *6th Rept., S. C. on Finance* [1817], p. 191.

49. William Snow Harris, *Our Dockyards. Past & Present State of Naval Construction* (Plymouth, 1863), pp. 6-11. This book was part of the campaign

that led to the creation of the Royal School of Naval Architecture and Marine Engineering in 1864.

50. ADM 1/3477, 27 May 1832.

51. *Martin Letters*, III, p. 393; ADM 7/577, pp. 111-16; H.C. (1837-38), xxxvii. 261-62.

Chapter 5: Winds of Change, 1834-1854.

1. *Rept. of the Select Committee on the Navy, Army and Ordnance Estimates*, p. xliii. H.C. (1847-48), xxi (i). 735.

Until 1838 some repairs were made at Portsmouth, since that yard already had the necessary equipment (used in conjunction with the mills, etc.). ADM 106/2293, 7 Nov. 1828; ADM 106/2296, 19 Aug. 1831.

2. ADM 1/5511, pt. 1, 1 April 1841; ADM 1/5537, 4 March 1844; ADM 1/5565, 31 Dec. 1846; BL, Add. MSS. 40450, fols. 122-25 (Herbert to Peel, 4 Sept. 1844), Add. MSS. 40556, fols. 11-12 (Herbert to Peel, 16 Dec. 1844); ADM 1/5716, 6 Jan. 1859, pp. 17-19; Edgar C. Smith, *A Short History of Naval and Marine Engineering* (Cambridge, 1937), pp. 74-75.

Paddle wheels had a number of disadvantages when used to propel naval vessels; however, it was because they took up space available in sailing ships for armament that they were not used for line-of-battle ships.

3. Printed in *Rept., Naval Yards, Appendix*, no. 16.

4. ADM 1/5564, 6 July 1846; H.C. (1847), xxxvi. 141-42; *Rept., S. C. on Navy Estimates*, p. lviii.

5. Fairbairn (one of the most brilliant engineers of his day and better known for his engineering works in Manchester) and Murray started the Millwall yard together in 1835 and there built several steam vessels and a number of engines for the Admiralty. The yard went bankrupt, and Murray became chief engineer at Portsmouth, in 1847.

6. *Rept., S. C. on Navy Estimates*, p. iv; ADM 1/5619, 19 Jan. 1853; *Rept. of the Committee on the Store Department*, p. 67. H.C. 1877, xxii. 67.

7. *Rept., Naval Yard*, Bromley's Report, pp. 489-90; Anon., *Review of the Course Pursued by the Shipbuilding Department of the Admiralty; between the Years 1832 and 1847* (Plymouth, 1847), pp. 77-79.

Bromley emphasized the importance of collecting the various departments of the Admiralty under one roof.

8. H.C. (1847-48), xli. 473-74.

9. ADM 1/5591, *Rept., Dockyard Revision*, fos. 2-3.

10. *Rept., Naval Yards*, Bromley's Report, p. 485; *Rept., S. C. on Navy Estimates*, pp. xliv-xlviii, li, liii; *Rept., S. C. on Navigation Laws, Evidence*, qq. 6265-68; ADM 1/5607, 19 June, 1850; ADM 83/54, No. 54.

11. *Rept., Naval Yards*, Bromley's Report, pp. 486-87; ADM 1/5591, *Rept., Dockyard Revision*, fos. 31-34.

12. The superintendents had been empowered since 1833 to have this done; none seems to have actually done so, however.

13. The son of Sir Thomas Byam Martin, Comptroller of the Navy, 1816-31.

14. ADM 1/5591, *Rept., Dockyard Revision*, fos. 31-24; BL, Add. MSS. 41434, fo. 115 (Martin to Admiralty, July 1857); *Repts., Dockyard Expenditure* [1886], p. 26; IND 12375, 41.20; H. C. (1860), xlii. 323.

15. H. C. (1847), xxxvi. 144.

16. *Ibid.*,142-43; *Rept., S. C. on Dockyard Appointments*, pp. iv, viii-ix, xiv; *Repts., Dockyard Expenditure* [1886], pp. 11, 19-20; ADM 1/6018, 7 June 1867.

17. ADM 1/5591, *Rept., Dockyard Revision*, fos. 12-13, 34-38.

18. ADM 1/5580, 27 Feb., p. 10; H. C. (1847), xxxvi. 148; ADM 1/5591, fos. 38-39; *Rept., Dockyard Economy*, pp. 50, 76-77, 81-82.

19. ADM 1/5591, *Rept., Dockyard Revision*, fos. 7, 33-35.

20. *Ibid.*, fo. 2.

21. *Ibid.*,fos. 2-4; *Rept., Naval Yards*, Bromley's Report, p. 488; *Rept., S. C. on Navy Estimates*, p. xlviii.

The superintendents's secretary, chosen from among the senior dockyard clerks and therefore bred in dockyard ways, was thought to be part of the problem. Henceforth, the superintendent chose his own secretary from outside the service. ADM 1/5607, 9 March 1850.

22. It will be recalled that the Commission on Fees had recommended regular and frequent surveys sixty years before.

23. ADM 1/5591, *Rept., Dockyard Revision*, fos. 5-7, 11-13; ADM 1/6108, 22 June 1869, *Rept., Committee on Store and Purchase Departments* (Devonport, pp. 3-4); *Rept., Naval Yards, Appendix*, no. 29 (General Order and Board Minutes, 25 Jan. 1849, pp. 536, 575-76).

24. ADM 1/5992, *Rept., Dockyard Revision*, fos. 9-11.

25. *Rept., Naval Yards*, p. 8, Bromley's Report, pp. 490-91.

26. This appears to be what Bromley originally preferred consistent with his proposal to create a comprehensive record for controlling dockyard expenditure.

27. ADM 1/5591, *Rept., Dockyard Revision*, fos. 14-17; *Rept., Naval Yards*, General Order, pp. 539-41, 576.

28. ADM 1/5591, *Rept., Dockyard Revision*, fos. 41-42; ADM 1/5567, 1 Dec. 1855.

29. ADM 1/5591, *Rept., Dockyard Revision*, fos. 19-22.

30. J. M. Haas, "The Best Investment Ever Made: The Royal Dockyard Schools, Technical Education, and the British Shipbuilding Industry, 1801-1914," *Mariner's Mirror*, 76 (1990), pp. 327-28; ADM 1/5537, 1 Feb. 1843; ADM 1/5582, 13 Jan. 1864; Adm 7/577, pp. 11-16; ADM 1/5585, 23 Jan. 1847, *Rept. on Dockyard Schools*, pp. 2, 6-21, 27, 34-36; ADM 1/5595, 5 May 1848, *Rept. on Apprentice Schools*.

31. Haas, "Best Investment," pp. 328-30; ADM 1/5595, 5 May 1848; ADM 1/5882, 13 Jan. 1864 (29 Nov. 1847); ADM 1/7824, Jan. 1905 (1 May 1904,

Rept. on Dockyard Schools); ADM 1/5742, 3 April 1860; ADM 1/6417, 24 March 1877, *Rept. of the Committee on the Royal Naval College*, p. xvii.

Chapter 6: Under Siege, 1854-1868.

1. ADM 1/5716, 6 Jan. 1859, *Rept. of the [Treasury] Committee on Navy Estimates from 1852 to 1858*, pp. 1-2, 10-11, 23-24, 28; ADM 1/5698, pt. 1, 20 May 1858, pt. 2, 16 Dec. 1858; *Rept., Dockyard Economy*, p. 153.

2. ADM 1/5716, 6 Jan. 1859, pp. 14-5, 17-19; H.C. (1860), xlii. 373.

3. *Rept., Dockyard Economy*, pp. 7, 98-99, 100-01.

4. *Ibid.*, pp. 181-88.

5. H.C. (1860), xliii. 287-95, 368, 383.

6. *Rept., Naval Yards*, pp. 5-6.

7. There was at first some disagreement over the relative merits of iron ships and composite (i.e. iron-clad wooden) ships.

8. [Duke of Somerset and Frederick Grey], The *Naval Expenditure from 1860 to 1866, and Its Results* (2nd edn., London, 1867), pp. 48-50, 82-83; Thomas Brassey, *The British Navy: Its Strengths, resources, and adminstration*, (5 vols., London, 1883), IV, pp. 114-15; Stanley Sandler, *The Emergence of Modern Capital Ship* (Newark, Del., 1979), p. 255; ADM 1/5913, 27 April 1865.

9. The journalist and naval correspondent Patrick Barry applauded the efficiency and enterprise of commercial yards and sneered at the dockyards. P. Barry, *Dockyard Economy and Naval Power* (London, 1863), pp. 209-312.

10. ADM 1/6082, 15 Oct. 1868. A visiting timber inspector was subsequently apppointed.

11. Robert Spencer Robinson (1809-89) had commanded various ships and was promoted Rear Admiral and appointed to the Royal Commission when his last ship was paid off in 1860. He was knighted in 1868 and promoted Admiral in 1871.

12. *Rept., Naval Yards, Evidence*, q. 53, Robinson's Memo., p. 586; ADM 1/5741, 16 Feb. 1860.

13. *Rept., Naval Yards*, Robinson's Memo., p. 605.

14. Graham told the Royal Commission that although the First Lord was technically not a minister of marine, the system could not work unless he was. Considerable controversy surrounded the constitutional relationship between the First Lord and the junior lords at this time and periodically down to the time of the First World War, when extracts from the relevent documents, going back to 1861, were collected. *Rept., Naval Yards, Appendix*, no. 10; ADM 116/3453; Murray, "Admiralty," pp. 463-70.

15. *Rept., Naval Yards*, pp. 6-8, Robinson's Memo., p. 605.

16. *Rept., Naval Yards*, p. 6, *Appendix*, no. 10; ADM 1/5853, 2 May 1863.

17. An interesting example of the influence of Thomas Arnold, whom the

Committee invoked.

18. A similar purpose had been served by the third Surveyor (1813-22).

19. *Rept., Dockyard Economy,* pp. 7-8, 33-34; ADM 1/5840, 4 Feb. 1863; ADM 1/5891, 30 Sept. 1864.

The dockyard instructions were revised by Arthur Price, a navy paymaster on half-pay, who had been secretary to the Royal Commission.

20. *Rept., Dockyard Economy,* pp. 19-20; H. C. (1860), xlii. 107; ADM 1/5742, 3 April 1860; ADM 1/5743, 16 Oct. 1861; ADM 1/5890, 14 June 1864; ADM 1/6018, 17 June 1867; *3rd Rept. of the Royal Commission on Labour, Evidence,* qq. 24, 405. H. C. (1894-95), xxii. 277; ADM 1/374, "Abstract of Petitions (1893)," p. 125.

21. *Rept., Naval Yards,* Robinson's Memo., p. 589; ADM 1/5892, 30 Nov. 1864; ADM 1/5936, 22 Dec. 1865; *Rept., Dockyard Economy,* p. 13.

22. ADM 1/5716, 6 Jan. 1859, p. 28; ADM 1/5765, 2 April 1861; ADM 1/5889, 5 March 1864; ADM 1/5936, 22 Dec. 1865; ADM 1/6232, 5 Feb. 1872; ADM 116/330, 14 Jan. 1891, pp. 274-79.

23. ADM 1/5591, fo. 45; ADM 1/5889, 5 March 1864; ADM 1/5936, 22 Dec. 1865.

24. The average day wage for all workers combined in 1865 was as follows: Royal Dockyards — 3s. 8d., Thames Ironworks — 4s. 6d., Millwall Ironworks — 4s. 5d. It was much less, however, in the provincial shipbuilding districts, e. g. just under 3s. at Napier's yard on the Clyde. Since the shortage of labor was serious only in the eastern yards, Robinson proposed raising wages there alone, the established men's just a little but the hired men's to what the competition was paying. ADM 1/5936, 22 Dec. 1865.

25. ADM 1/5667, 8 Dec. 1855; IND 12439 (1857), 41.14, 41.8; BL, Add. MSS. 41434, fos. 115-17 (Martin to Admiralty, July 1857); IND 12455 (1858), 41.12, 41.9; IND 12487 (1860), 41.13; *Rept., Dockyard Economy,* p. 29; *Rept., Naval Yards,* p. 10, Robinson's Memo., p. 594; ADM 1/5841, 13 June 1863; ADM 1/5889, 5 March 1864; ADM 1/5890, 20 May, 1 June, 24 June (9 and 14 June) 1864; ADM 1/5936, 22 Dec. 1865.

Unskilled laborers got a small rise in 1861 (although much less than the Dockyard Committee had recommended) and a larger rise in 1867 (on the latter occasion because of a plea from Portsmouth corporation out of concern that otherwise they would become a charge on the poor rates), but were still paid starvation wages of 18s. per week. Why, it was argued, should they be paid more when even at such wages supply always exceeded demand? *Rept., Dockyard Economy,* pp. 35, 58; ADM 1/5975, 25 May 1866; ADM 1/6018, 18 June 1867.

26. ADM 1/5841, 1 May and 13 June 1863; ADM 1/5885, 30 Sept. 1864; ADM 1/5890, 24 June 1864.

27. Employment in shipbuilding and marine engineering on the Thames fell from 27,000 in 1865 to 9000 in 1871 and to 5500 in 1901. S. Pollard, "The

Decline of Shipbuilding on the Thames," *Econ. Hist. Rev., 2,* III (1950), pp. 79, 88.

28. ADM 1/6079, 11 Feb. 1868; ADM 1/6016, 10 March 1869; ADM 1/6211, 7 March 1871.

29. The Dockyard Committee and the Royal Commission both recommended reviving task work for new construction and major repairs; that was not on the cards, though; Seppings's new scheme, drawn up on Melville's instructions in 1830-31, had been tried in 1847 (on Bromley's advice) and found to be just as unsatisfactory as the original scheme of 1775. The Dockyard Committee said the problem was with the master shipwrights; they did not know how to carry task work out and should learn from commercial builders who did. *Rept., Dockyard Economy,* pp. 31-33, 115-18, *Evidence,* qq. 2642-2643; *Rept., Naval Yards,* p. 11, *Evidence,* qq. 6827-6855, *Appendix,* no. 34; H. C. (1860), xlii. 319, 324, 389-90.

30. *Rept., Dockyard Economy,* pp. 25, 28, 33, 110-15, 153, 197-201; *Rept., Naval Yards, Evidence,* qq. 2857-2861, Robinson's Memo., p. 594; ADM 1/5667, 8 Dec. 1855; IND 12455 (1858), 41.12; H. C. (1860), xlii. 304, 324.

31. *Rept., Dockyard Economy,* pp. 23-25; *Rept., Naval Yards, Appendix,* no. 47, Bromley's Report, pp. 540-41, Robinson's Memo., p. 597.

32. *Rept., Naval Yards, Evidence,* qq. 2865-2867, *Appendix,* no. 41b; *A Scheme for Prices for Performing Shipwrights' Work* (London, 1857); *A Scheme of Prices for Performing Smiths' and Millwrights' Work* (London, 1857).

There were more than 4800 articles of shipwrights' work but separate rates for each class of ship.

33. *Rept., Naval Yards,* p.11.

34. *Rept. of the Committee on Dockyard Manufactures,* p. 23. H.C. (1865), xxv. 23.

35. ADM 1/3462, Z.

36. ADM 1/5889, 6 Feb. 1864 (17 Feb. 1864); ADM 1/5931, 18 Jan. 1865; *Rept. from the Accountant General of the Navy,* p. 194. H.C. (1865), xxxv (1).

37. Quoted in Solomons, p. 19.

38. *Rept., Naval Yards,* p. 8, *Evidence,* qq. 827-836, 2806-2811, 6833-6855, 8313-8364, 8505, Robinson's Memo., pp. 588, 596-97; Barnaby, pp. 313, 435; ADM 1/5977, 22 Nov. 1866 (26 Nov. 1866); Parkinson, p. 177.

39. *Rept., Naval Yards,* p. 11.

40. The Controller by 1860 was receiving 10,000 returns a year. *Rept., Naval Yards, Appendix,* no. 31.

41. These two offices were combined at the four smaller dockyards.

42. H.C. (1860), xlii. 330-32; *Rept., Naval Yards,* pp. 8-10.

43. The title had formerly been Secretary of the Admiralty.

44. Childers was particularly well suited to continue and complete the work begun by Bromley. He had been Auditor General of Victoria (1852-53), was a director of the London and County Bank, and had recently acted for Baring

Brothers in connection with a large Australian loan.

45. *Ibid.*, p. 10; *Rept., Dockyard Manufactures*, pp. 1-2.

46. *Rept. from Accountant General*, pp. 169-70; ADM 1/5765, 14 June 1861; ADM 1/5836, 27 Feb. 1863; ADM 1/5913, 15 March 1864 (1 April 1864, pp. 12-13); *4th Rept. of the Committee of Public Accounts* (1889), in *Epitome of the Reports from the Committee of Public Accounts*, p. 240. H.C. (1937-38), xiii. 240.

47. ADM 1/6080, 28 May 1868 (9 March 1868).

48. ADM 1/5889, 23 March 1864; ADM 1/5883, 29 April 1864; ADM 1/6081, 25 Aug. 1868; ADM 1/5977, 22 Nov. 1866 (26 Nov. 1866); ADM 1/5935, 28 Aug. 1865 (30 June and 22 July 1865); ADM 1/6079, 19 and 27 Feb. 1868; *Rept. from Accountant General*, pp. 137-40, 171-72, 191-92, 206, 261-63.

49. *Rept., Committee on Contracts*, p. 1.

50. *Rept. from Accountant General*, pp. 191-92, 195-220; ADM 1/5932, 3 March 1865 (Dockyard Circular, 1 March 1865); ADM 1/5932, 3 March 1865 (*1st Rept. of the Committee on Dockyard Books and Forms*, 25 Feb. 1865 *passim*); ADM 1/5941, 7 April 1865; ADM 1/5961, 31 March 1865; ADM 1/5943, 3 Nov. 1865.

51. Roseveare, p. 139.

52. Palmerston to Gladstone, 21 Sept. 1865, quoted in Henry Parris, *Constitutional Bureaucracy. The Development of British Central Administration since the Eighteenth Century* (London, 1969), p. 125.

53. *Rept. of the Select Committee on Admiralty Monies and Accounts*, pp. xx-xxi. H.C. (1867-68), vi; ADM 1/6020, 18 Dec. 1867 (20 Sept. 1867, "Examination of Cost of Ships of 'Amazon' Class," pp. 2-3, 15, 17-18, 21-25); ADM 1/6079, 19 and 27 Feb. 1868; ADM 1/6080, 26 May 1868, 28 May 1868 (9 March 1868).

54. ADM 1/6080, 27 May 1868; *Rept., Admiralty Monies*, pp. xv-xviii.

55. This was done by Frank Fellowes, Inspector of Yard Accounts.

56. ADM 1/6062, 16 Oct. 1868; ADM 1/6157, 23 Feb. 1870; ADM 1/6492, 1 Jan. 1879, fols. 1-10.

57. Platers prepared plates, angle bars, etc. by marking, shearing, rolling, bending, shaping, etc. Each plater was assisted by a gang of four to six helpers.

58. Holders-up assisted the riveters by hammering the rivets into their proper position in the holes in the plates the riveters were clinching together.

59. J.E. Mortimer, *History of the Boilermakers' Society*, Vol. I, *1834-1906* (London, 1973), pp. 27-29; Abel, pp. 110-12.

60. ADM 116/330, 4 Dec. 1890, pp. 16-17, 45-46; Hansard, 4, ix, col. 1141 (6 March 1893); *3rd Rept., Royal Commission on Labour, Digest of Evidence*, p. 58.

61. Mortimer, pp. 60-64, 160.

62. Scott Russell was one of the greatest shipbuilders, and the *Great Eastern* was the biggest merchant vessel, of the time.

63. *Rept. of the Select Committee of the House of Lords on the Board of Admiralty*, p. 56. H.C. (1871), vii. 56; ADM 1/5890, 18 April, 1866.

64. ADM 1/5889, 23 March 1864.

65. Barnaby was Reed's brother-in-law and succeeded him as chief constructor in 1871.

66. Haas, "Best Investment," pp. 327-30, 333; Pollard and Robertson, p. 144.

Chapter 7: Revolution and Counterrevolution, 1868-1885.

1. Woolwich continued to be the principal store depot; Deptford yard was subsequently sold.

2. ADM 1/6079, 11 Feb. 1868; ADM 1/6106, 10 March 1869; ADM 1/6159, 9 Aug. 1870 (4 Aug. 1870); ADM 1/6232, 15 Feb. 1872; ADM 116/330, 14 Jan. 1891, pp. 274-79.

The Admiralty in 1869 assisted the emigration to New Brunswick of about 300 redundant dockyardmen and their families in a scheme that included the royal arsenals. A total of 1189 men, women, and children went out. ADM 1/6107, 1 May 1869; ADM 1/6108, 8 June 1869.

3. ADM 1/6081, 30 July 1868; ADM 1/6183, 23 June 1870.

4. "Childers was responsible [when Financial Secretary of the Treasury (1865-66)] in introducing a new class of writers into the Customs Department; shortly afterwards the Admiralty adopted a similar idea. The significance of these two experiments was their anticipation of the general writer class introduced into the whole [Civil] Service in 1870." Maurice Wright, *Treasury Control of the Civil Service 1854-1874* (Oxford, 1969), pp. 36, 182, 186.

5. *Rept., S.C. on Board of Admiralty, Evidence*, q. 447.

6. *Rept., Megaera*, pp. 23-24.

7. *Rept., Megaera, Evidence*, q. 14,907.

8. *Rept., S.C. Board of Admiralty*, p. 6.

9. Quoted in Spencer Childers, *The Life and Correspondence of the Right Hon. Hugh C.E. Childers 1827-1901* (2 vols., London, 1901), I, p. 161.

10. ADM 1/6062, 23 Dec. 1868; ADM 116/861, 2 Jan. 1869; H.C. (1868-69), xxxviii. 314-16; Murray, "Admiralty," pp. 471-75.

11. *Rept. of the Committee on the Store Department*, p. 162. H.C. (1877), xxii. 162.

12. *Rept., S.C. Board of Admiralty, Evidence*, qq. 482-485; ADM 1/6104, 16 Jan. 1869; ADM 1/6106, 3 March 1869; ADM 1/6107, 3 May 1869 (20 April 1869, pp. 1-2); ADM 1/6311, 22 July 1875 (instructions issued in 1870).

Naval ordnance (split off from the War Department in 1866) also was put under the Controller.

13. ADM 1/6104, 7 Jan. 1869; ADM 1/6105, 11 Feb. 1869.

14. *Rept., Store Department, Evidence*, q. 2639, *Appendix,* no. 22; ADM 1/6104, 16 Jan. 1869 (Robinson's Proposals, 8 Feb. 1869, fos. 1-4); ADM

1/6157, 23 Feb. 1870; ADM 1/6656, 24 June 1882 ("Alterations in Store Department," 31 Dec. 1883, p.1).

15. This was in 1876. An intentional slight, perhaps, to his successor but one, Rear Admiral Houston Stewart? Stewart between 1863 and 1872 had successively been superintendent of Chatham, Devonport and Chatham yards.

16. *Rept., Store Department, Evidence*, qq. 2619-2639, *Appendix*, no. 50.

17. The storekeeper's and cashier's departments were combined at Sheerness and Pembroke.

18. ADM 1/6114, 21 Dec. 1869, fos. 5-11; ADM 1/6157, 9 and 23 Feb. 1870; *Rept., Store Department, Appendix*, no. 22.

19. Winston Churchill held only twenty-four boards in 1913 and only seventeen in 1914. Arthur J. Marder, *From Dreadnought to Scapa Flow: The Royal Navy in the Fisher Era, 1904-1919* (5 vols., London, 1961-79), II, p. 302.

20. *Rept., S.C. Board of Admiralty, Evidence*, q. 348..

21. Sandler, ch. 7.

22. The Admiralty had been trying since 1859 to get private firms to design ships for the Royal Navy; the design for the *Captain* was the only one approved, however. ADM 1/6970, 8 Jan. 1889, pp. 1-2.

23. Reed became chairman and general manager of Earle's Shipbuilding and Engineering Co., Ltd., of Hull, which specialized in warships. He entered Parliament as a Liberal in 1874, was knighted in 1880, and was a Lord of the Treasury in Gladstone's third administration (1886).

24. ADM 1/6157, 9 Feb. 1870.

25. *Rept., Store Department, Evidence*, qq. 2653-2658.

26. N.A.M. Rodger, "The Dark Ages of the Admiralty, 1869-85. Part I: 'Business Methods', 1869-74," *Mariner's Mirror*, LXI (1975), p. 338.

27. *Rept., S.C. on Board of Admiralty*, pp. ix-xiii; ADM 1/6309 (1), 19 Feb. 1874.

For a recent criticism of Childers's adoption of "business methods" and "personal responsibility", see Rodger, "Dark Ages," p. 338.

28. *Rept., S.C. on Board of Admiralty*, p. ix; ADM 1/6309 (1), 19 Feb. 1874; Murray, "Admiralty," pp. 475-76.

29. ADM 1/6658, 10 March 1882.

The debate over the Controller had broken out again in 1876 with Childers and Robinson on one side and the Controller, Rear Admiral Houston Stewart, on the other. *Rept., Store Department, Evidence*, qq. 502, 2697-2705, *Appendix*, no. 50.

30. The title was orginally Chief Naval Architect but was changed to Director of Naval Construction in 1875.

31. Manning, p. 89; H.C. (1872), xxxix. 340-41; ADM 1/6183, 19 March 1872; ADM 1/6235 (2), 29 July 1872; *Rept. of the Committee on the Entry, Training, and Promotion of the Professional Officers of the Dockyards, Evidence*, q. 714. H.C. (1883), xvii. 70.

32. Barnaby, Barnes and Reed had been contemporaries at the second School of Naval Architecture in the 1850s. Reed was married to Barnaby's sister.

33. ADM 1/6235 (2), 29 July 1872; ADM 1/6183, 9 Aug. 1872; ADM 1/6236 (1), 8 Aug. 1872 and 2 Jan. 1873; ADM 1/6293, 5 Feb. 1873; ADM 116/155 (Stewart to Admiralty, 2 Jan. 1875); *Rept., Store Department, Evidence*, q. 855.

34. ADM 116/155.

35. ADM 1/6274 (2), 1 Nov. 1873; ADM 1/6293, 12 Dec. 1873; ADM 116/205; ADM 116/330, 4 Dec. 1890, p. 46.

36. This was because the question of the master shipwright's department had to be settled first.

37. ADM 1/6438, 11 July 1877; ADM 1/6492, 1 Jan. 1879 (11 May 1878, fos. 4-5); ADM 1/6541, 21 Aug. 1880; *Rept., Dockyard Economy*, p. 10; *Rept., Professional Officers*, pp. 8-9; *Repts., Dockyard Expenditure* [1886], pp. 10, 19, 22, *Evidence*, qq. 1245a-1259, 1434-1442, 2291-2295.

38. Hansard, 3, ccxiv, cols. 938-39; 943, 945-46 (25 Feb. 1873); ADM 116/155.

The salaries of the chief engineers at Portsmouth and Devonport were raised to a maximum of £650 per annum.

39. McHardy was appointed Superintendent of Stores in 1869 on Robinson's recommendation. He had until then been Robinson's secretary.

40. *Rept., Store Department*, pp. iii-iv, *Evidence*, qq. 152-178, 482-485, *Appendix*, no. 26 (pt. 13).

41. *Rept., Store Department*, pp. xii, xiv-xvi, *Evidence*, qq. 110-113, 2650-2655, *Appendix*, nos. 7, 9, 22, 26 (pts. 1 and 2).

42. *Repts., Dockyard Expenditure* [1886], *Evidence*, qq. 1097-1101, 1106-1110, 2231-2233.

43. Hansard, 3, ccciv, cols. 938-39, 943 (23 Feb. 1873); ADM 1/6541, 2 Aug. 1880; *Repts., Dockyard Expenditure* [1886], p. 9; ADM 1/6389, 28 April 1876.

44. ADM 116/231; ADM 1/6588, 2 Nov. 1881.

45. Brassey, IV, p. 23.

46. Another 327 (20 per cent) were dealt with locally.

47. ADM 1/6105, 9 Feb. 1869; ADM 1/6492, 1 Jan. 1879 (11 May 1878, fos. 8-9); *Rept., Professional Officers, Evidence*, qq. 30-52, 649-654. 863-887; ADM 1/7594c (29 Jan. 1902), *Second Rept. of the Committee on the Controller's Department, Appendix*, no. 2.

48. ADM 1/6492, 1 Jan. 1879 (11 May 1878, fos. 6-8).

49. ADM 1/6104, 30 Jan. 1869; ADM 1/6157, 23 Feb. 1870, pp 3-4; *Rept., Store Department*, pp. x-xi; ADM 1/6108, 22 June 1869 (Devonport, pp. 3-4; Portsmouth, pp. 1-3; Chatham, pp. 1-3; Memo. of Instructions, pp. 1-2, 7-8 [Appendix C, pp. 1-4]).

50. ADM 1/6108, 22 June 1869 (Memo. of Instructions, pp. 3-4); *Rept., Store*

Department, p. v; ADM 1/6656, 24 June 1882, p. 1; *Rept., Dockyard Expenditure* [1886], p. 67.

51. ADM 1/6108, 22 June 1869 (Memo. of Instructions, p. 1); ADM 1/6656, 24 June 1882, p. 2.

There had at first been considerable difficulty in procuring iron of the requisite quality and having it delivered promptly. This led in 1864 to a proposal, strongly supported by Robinson, to purchase the Atlas Steel and Iron Works of Sheffield from John Brown & Co., who were prepared to sell for £750,000. The Board of Admiralty, however, did not approve. The problem appears to have largely been worked out by the 1870s. ADM 1/5841, 29 May 1863; ADM 1/5889, 17 Feb. 1864; *Rept., Store Department, Appendix,* no. 26 (pt. 2).

52. ADM 1/6273 (1), 9 Jene 1873; *Rept., Store Department,* p.v; ADM 1/6656, 24 June 1882, pp. 2-3; *Repts., Dockyard Expenditure* [1887], p. 373.

53. *Rept., Store Department,* p. ix, *Evidence,* q. 211; ADM 1/6390, 13 Oct. 1876, paras. 17-19, 22; ADM 1/6656, 24 June 1882, pp. 3, 12-13.

54. *Ibid.,* pp. 3-5; ADM 1/6679 (1), 6 June 1883.

55. ADM 1/6656, 24 June 1882, pp. 7, 12; ADM 116/223, p. 3; ADM 1/6774, Northbrook's Minute, 2 July 1885, p. 51; *Repts., Dockyard Expenditure* [1886], *Evidence,* qq. 2276-2278.

56. Starting with Portsmouth.

57. ADM 116/233; ADM 116/238, pp. 5-38 *passim;* ADM 1/6774, 2 July 1885, p. 52; ADM 1/6492, 1 Jan. 1879 (11 May 1878), fos. 11-13.

58. ADM 1/6539, 7 March 1880 (*Rept., of the Committee on the Reorganization of the Dockyard Storehouse Staff,* pp. 3-8, 14-15, Order of 29 April 1880, p. 3).

59. C. M. Palmer, M. P. — Palmer's Shipbuilding Co.; John Burns (later Lord Inverclyde) — Cunard Steamship Co.; Thomas Ismay — White Star Line; Joseph Samuda — Samuda Brothers, shipbuilders.

60. *Rept., Committee on Contracts,* pp. 1-3.

61. It was, significantly, in 1884 that Armstrong-Mitchell commenced building their new warship yard at Elswick.

62. *Repts., Dockyard Expenditure* [1886], p. 70; ADM 1/7114, 13 Jan. 1892 ("Comparative Cost of New Construction," p. 8).

63. *Ibid.,* p. 6.

64. *Repts., Dockyard Expenditure* [1886], *Evidence, passim*; ADM 1/7114, 3 Jan. 1892 (Cost of Construction", p. 6).

65. However, warship contractors complained to Ravensworth's committee about "the incomplete and meagre character of the specifications furnished by the Admiralty." *Rept., Committee on Contracts,* p. 1.

66. *Repts., Dockyard Expenditure* [1886], pp. 24-25, *Evidence,* qq. 2242-2248.

Private yards also filled out much of the detail of the plans as the construction of merchant vessels advanced. Parkinson, p. 118.

67. ADM 116/36 (Elgar's memo., 21 Aug. 1889).

68. *Repts., Dockyard Expenditure* [1887], p. 379, *Evidence,* qq. 1157, 1308-1311, 2231-2233; *Repts., Dockyard Expenditure* [1886], *Evidence,* qq. 1157, 1308-1311, 2231-2233; Pollard and Robertson, p. 214.

69. *Repts., Dockyard Expenditure* [1886], pp. 20-21, *Evidence,* qq. 1207-1215, 2291-2295, 2508-2546, " Remarks by Mr. Barnes," pp. 142-143.

70. The other members were G.W. Rendel (Additional Civil Lord), Adm. Sir G.T.P. Hornby (President of the Royal Naval College), and Nathaniel Barnaby (Director of Naval Construction).

71. *Rept., Dockyard Economy,* pp. 11-12; *Rept., Professional Officers, Evidence,* q. 724; Brassey, IV, p. 51.

72. They had never been very detailed; it was not necessary in the age of wooden shipbuilding that they should be. Ship design and construction had, however, become vastly more complex in the last twenty five years and the designer needed to provide the builder with more complete drawings and specifications.

73. *Rept., Professional Officers,* p. 16, *Evidence,* qq. 149, 230, *Appendix,* no. I; *Repts., Dockyard Expenditure* [1886], p. 24; ADM 1/8220, *Report of the Committee on the Royal Corps of Naval Constructors,* p. 69.

74. *Rept., Professional Officers,* pp. 9-10, *Evidence,* qq. 158-65, 486-490.

75. *Rept., Professional Officers,* p. 8, *Evidence,* qq. 161, 230-239, *Appendix,* no. I; ADM 1/5742, 3 April 1860.

76. *Rept., Professional Officers, Evidence,* qq. 230-239, 540-546, *Appendix,* no. I.

77. *Rept., Professional Officers, Evidence,* qq. 500-556.

78. *Rept., Professional Officers,* p. 10, *Appendix,* no. I; ADM 1/8220, *Rept., Royal Corps,* p. 7; ADM 1/6772, 14 March 1885; ADM 1/6774, 2 July 1885; ADM 1/6870, 1 July 1887.

79. *Rept., Professional Officers,* pp. 12-15; ADM 1/8220, *Rept., Royal Corps,* pp. 5, 7; ADM 1/6772, 14 March 1885.

80. *Ibid.*

Chater 8: The Watershed, 1885-1900.

1. Three subcommittees — on dockyard accounts and audit, revision of dockyard forms and returns, and supplies of stores and materials — reported separately.

2. Graham had previously been superintendent at Malta. He was knighted in 1887 and became President of the Royal Naval College (1888-9) upon retiring from the Controllership.

3. Lord George Hamilton's "Introduction" to Manning, pp. vii-viii.

4. Hansard, 4, cccxxxii, col. 1171 (7 March 1889).
The two-power standard had tacitly been adhered to for most of the century.

5. Spencer was First Lord from 1892 to 1895. He was succeeded by G.J. Goschen (1895-1900).

6. Sumida, Table 6, "Expenditure under Vote 8 (Shipbuilding, Repairs, Maintenance, etc.) and Vote 9 (Naval Armaments) of Navy Estimate, 1889-90 to 1913-14."

7. *Repts., Dockyard Expenditure* [1886], pp. 13, 25.

8. *Repts., Dockyard Expenditure* [1886], pp. 20-21, 146.

9. ADM 1/6971, 12 March 1889; *Repts., Dockyard Expenditure* [1887], pp. 337-38.

10. *Repts., Dockyard Expenditure* [1886], p. 146.

11. It has been suggested that Elgar's appointment was directed as much against White as it was intended for the benefit of the service. Pollard and Robertson, p. 209.

12. The exchange of views between White and Ripon is printed in Manning, pp. 195-204.

13. White to Hamilton, Aug. 1891, quoted in *ibid.*, pp. 283-85.

14. *Repts., Dockyard Expenditure* [1886], pp. 16, 24-25, *Evidence*, qq. 1525-1534.

15. *Repts., Dockyard Expenditure* [1886], pp. 16, 25.

16. ADM 1/6816, 12 Jan. 1886; ADM 1/7807, 5 Jan. 1905, *Rept. of the Committee on Naval Establishments*, p. 11.

17. *Repts., Dockyard Expenditure* [1886], pp. 21-22; ADM 1/6926B, 4 Dec. 1888; ADM 1/7158, 24 July 1893.

18. *Repts., Dockyard Expenditure* [1886], pp. 21-22; ADM 1/6926B, 14 Jan. 1888; ADM 1/7158, 24 July 1893; ADM 116/358.

19. *Repts., Dockyard Expenditure* [1886], p. 150; *4th Rept., Public Accounts Committee* [1889], pp. 237-38.

20. The old office of the same name was abolished in 1877.

21. Waterhouse was a partner in the well-known firm of accountants bearing his name. Mills was the Treasury's officer of accounts.

22. ADM 1/6868B, 17 March 1887; ADM 1/7990, 14 Jan. 1908, *3rd Rept. of the Subcommittee on Accounts*, pp. 3-4; *4th Rept., Public Accounts Committee* [1889], p. 237.

Another, but similar, plan dividing the yard accounting staff between the chief constructor and the storekeeper was originally adopted in 1886 and abandoned in 1887 when Elgar objected that it would interfere unduly with the chief constructors' duties. *Repts., Dockyard Expenditure* [1886], p. 161, *Evidence*, qq. 158, 175-176; *Repts., Dockyard Expenditure* [1887], pp. 358-59, *Evidence*, qq. 3178-3203.

23. *4th Rept., Public Accounts Committee* [1889], pp. 238-40; ADM 1/7990, 14 Jan. 1908, *3rd Rept., Subcommittee on Accounts*, p. 4.; *3rd Rept., Public Accounts Committee*, p. 17. H.C. (1890-91), xi. 117.

Several of the country's largest shipbuilding and engineering firms in 1907 had a practically identical system of independently recording work. It is impossible to say, however, whether they or the Admiralty were first. ADM 1/7990, 14 Jan. 1908, *3rd Rept., Subcommittee on Accounts*, p. 4.

24. *4th Rept., Public Accounts Committee* [1889], pp. 187-90, 242-43; ADM 1/7990, 14 Jan. 1908, *3rd Rept., Subcommittee on Accounts*, p. 13.

25. *Ibid.*, pp. 14-20.

26. *4th Rept., Public Accounts Committee* [1889], pp. 187-90, 242-43; ADM 1/7990, 14 Jan. 1908, *3rd Rept., Subcommittee on Accounts*, p.15.

27. *Ibid.*, pp. 3-4; William Ashworth, "Economic Aspects of Late Victorian Naval Administration," *Econ. Hist. Rev.*,2, xxii (1969), pp. 501-02, 504.

28. ADM 116/330 (Memo. by Elgar, Feb. 1891; remark by Forwood, p. 6; Observations on Rept., Dockyard Economy, pp. 46-48).

29. *Repts., Dockyard Expenditure* [1886], pp. 7-18, *Evidence*, qq. 1167-1178, 2276-2278, 2307, 2560, *Appendix*, Instructions, p. 174; Eustace Tennyson D'Eyncourt, *A Shipbuilder's Yarn. The Record of a Naval Constructor* (London, 1948), pp. 29 ff.; *3rd Rept., Royal Commission on Labour, Digest of Evidence*, p. 5; ADM 116/330, 14 Jan. 1891, pp. 274-79; ADM 1/7465C, 4 Dec. 1900 (16 Sept. 1897).

30. This is a good example of instructions that forced upon officials duties which they could not perform.

31. ADM 1/6588, 2 Nov. 1881, *Dockyard Regulations* (1882), p. 28; *Repts., Dockyard Expenditure* [1886], *Evidence*, qq. 1245a-1259, 2560; *Rept., Dockyard Expenditure* [1887], pp. 342-43; *3rd Rept., Royal Commission on Labour, Digest of Evidence*, p. 59.

32. ADM 116/84 *passim*

33. ADM 116/316, 23 July 1889; ADM 116/84 *passim*; ADM 1/7465C, 4 Dec. 1900 (16 Sept. 1897).

34. ADM 116/1028, 19 Oct. 1906.

35. ADM 1/7661, 20 1903; Hansard, 4, cxxxii, col. 571 (23 March 1904); ADM 1/7114, 3 Jan. 1892 ("Cost of Construction," p. 5).

36. Lord George Hamilton, *Parliamentary Reminiscences and Reflections 1886-1906* (London, 1922), p. 100; Hansard, 4, cccxxiii, cols. 1170, 1175 (2 March 1889); ADM 6/330 (Elgar's memo., Feb. 1891); Parkes, pp. 293-338 *passim*.

37. She was laid down at Pembroke in March 1886 and completed at Portsmouth in July 1891.

38. The Admiralty as a matter of policy never disclosed the reason for differences in the cost of similar ships built in different yards, not even to the Comptroller and the Auditor General or the Committee of Public Accounts. *1st Rept., Committee of Public Accounts* [1896], in *Epitome of Repts., Committee of Public Accounts*, pp. 376-77.

39. ADM 1/7114, 13 Jan. 1892 "Cost of Construction," pp. 3-6, 8; ADM 6/330 (Elgar's memo., Feb. 1891).

40. Sumida, pp. 15, 17-18; Bernard Mallett, *British Budgets, 1877-78 to 1912-13* (London, 1913), pp. 49-52; Parkes, pp. 393-415 *passim*; ADM 1/7520, 25 April 1901, *Rept. of the Committee on Arrears of Shipbuilding* [1902], pp. 4-16, 33-66.

41. Based on costs as stated in Parkes, pp. 358, 381, 393.

42. ADM 1/8046, 22 April 1909 (C. McLaren to R. McKenna, 2 April 1909 and covering letter of 22 April 1909); *Rept. of the Select Committee on Estimates, Evidence*, qq. 1182, 1191. H.C. (1913), vi. 315; Hugh Lyon, "The relations between the Admiralty and private industry in the development of warships," in Bryan Ranft, ed., *Technical Change and British Naval Policy 1860-1939* (London, 1977), pp. 60-62; ADM 1/7114, 13 Jan. 1892 ("Cost of Construction," p. 5).

43. Hamilton, p. 82; Pollard and Robertson, pp. 115-29.

44. ADM 1/7114, 13 Jan. 1892 ("Cost of Construction," pp. 6-7); ADM 1/7520, 25 April 1901, *Rept., Arrears of Shipbuilding*, p. 12.

45. Ripon's memorandum, 8 March 1886, quoted in Manning, pp. 195, 206; ADM 116/316.

46. Manning, p. 192. What White actually recommended was that the use of piecework, at that time quite limited, should be extended.

47. ADM 116/330, 4 Dec. 1890, pp. 21-22; *3rd Rept., Public Accounts Committee* [1890-91], pp. 117-18, *Evidence*, q. 1419; *3rd Rept., Royal Commission on Labour, Evidence*, qq. 25,913-25,920, *Digest of Evidence*, pp. 31, 33, 38, 40, 50; ADM 116/1067, 2 March 1912; ADM 1/7465C, 4 Dec. 1900 (Paget's Report, 6 Sept. 1897); Pollard and Robertson, pp. 182-84.

The Thames Ironworks in 1890 adopted a scheme that seems to have been slightly similar to the dockyards' scheme of tonnage payment. *3rd Rept., Royal Commission on Labour, Evidence*, qq. 24,907, 25,809.

48. *Repts., Dockyard Expenditure* [1887], p. 335.

49. *Repts., Dockyard Expenditure* [1886], pp. 19-20, *Evidence*, qq. 1128-1133.

50. See Haas, "Trouble at the Workplace," pp. 210-19.

51. The Associated Shipwright' Society had absorbed the rival dockyards' Ship Constructive Association and a number of smaller unions representing various trades. Clegg, pp. 152 and n. 5, 443 and n. 2; *3rd Rept., Royal Commission on Labour, Digest of Evidence*, p. 590.

52. ADM 116/330, 4 Dec. 1890, p. 5; *3rd Rept., Royal Commission on Labour, Evidence*, qq. 24,053-24,055, *Digest of Evidence*, p. 57.

53. Hansard, 4, ix, col. 1141 (6 March 1893); ADM 116/330, 4 Dec. 1890, pp. 17, 46.

54. ADM 116/330 (4 Dec. 1890, pp. 6-17, 4-42, 91-100; 14 Jan. 1891); ADM 116/316, 23 July 1889; ADM 116/374, Abstract of Petitions, May 1893, p. 45; *3rd Rept., Royal Commission on Labour, Evidence*, q. 24,408, *Digest of Evidence*, p. 60; ADM 1/7114, 7 Jan. 1892.

55. ADM 116/374, 22 July 1893, p. 10; ADM 116/330, 4 Dec. 1890, pp. 16-17, 44-45, 175-78; *3rd Rept.*, *Royal Commission on Labour, Digest of Evidence*, pp. 28, 60; B. Seebohm Rowntree, *Poverty: A Study of Town Life* (2nd ed., London, 1902), p. 110.

56. ADM 116/330, 4 Dec. 1890, pp. 5-6.

57. ADM 116/330, 4 Jan. 1891, p. 23; T. Good, "Shipbuilding Labour Troubles," *World's Work,* XVII (1913), p. 609; Hansard, 4, ix, col. 1143 (6 March 1893).

58. ADM 116/330, 4 Dec. 1890, pp. 5-7, 9.

59. *Ibid.,* p. 7, 10, 13; ADM 1/6973, 4 June 1889 (5 Aug. 1890); ADM 116/374, Abstract of Petitions, May 1893, p. 3; *3rd Rept., Royal Commission on Labour, Digest of Evidence,* p. 57.

60. ADM 116/330, 4 Dec. 1890, pp. 25-26; ADM 1/6975, 8 Oct. 1891; H.C. (1890-91), liii. 368-70; ADM 1/7206, 3 Jan. 1894.

61. ADM 116/330 (4 Dec. 1890, pp. 7-8; 13 March 1891; 14 March 1891; 25 March 1891); ADM 116/374 (23 Aug. 1893; Abstracts of Petitions, May 1893, pp. 45, 77); Hansard, 4, ix, cols. 1138-39 (6 March 1893); *3rd Rept., Royal Commission on Labour, Evidence,* q. 24,388; *Repts., Dockyard Expenditure* [1886], *Evidence,* q. 1304.

It is perhaps significant that 19 per cent of the shipwrights (apparently the best) did not object to classification.

62. ADM 116/374, Abstract of Petitions, May 1893; Hansard, 4, ix, cols. 1129-80 (6 March 1893).

63. ADM 116/374, (30 June 1893, 22 July 1893; Treasury to Admiralty, 26 Aug. 1893; Admiralty to Treasury, 4 Sept. 1893); ADM 1/7874, 24 March 1906 (Marshall, "Remarks on points raised in federation minutes").

64. ADM 116/374, Abstract of Petitions, May 1893, p. 2; ADM 1/7206, 3 Jan. 1894 (20 Dec. 1893, 3 Jan. 1894); J. Harris, *Unemployment and Politics: A Study in English Policy, 1886-1914* (Oxford, 1972), pp. 66, 69; Good, p. 611.

65. H.C. (1896), liv. 542-2.

Chapter 9: A New Century and New Ideas, 1900-1914.

1. Of the two First Lords between Cawdor and Churchill, the Earl of Tweedmouth (1905-08) was weak and going mad, and Reginald McKenna (1908-11), although a good administrator, left no very great mark on the management of the dockyards.

2. The Committee consisted of the Second Sea Lord (Rear Admiral Archibald Douglas), the Permanent Secretary (Sir Evan McGregor) and the Accountant General (Richard Ormsby).

3. ADM 1/7807, 5 Jan. 1905 (Selborne's letter, 5 Jan. 1905); ADM 1/7594c, 29 Jan. 1902, *2nd Rept. of the Committee on the Controller's Department,* fos. 6-7; ADM 1/7530, *Rept. of the Committee on the Controller's Department.*

4. Not all of these emanated from the Controller's department, however.

5. ADM 1/7520, 30 April 1891; ADM 1/7524, 19 July 1901; ADM 1/7594c, 29 Jan. 1902, *Rept., Controller's Department*, fos. 1-4; ADM 1/7617, 26 Nov. 1902, *Rept. of the Committee on the Department of the Director of Naval Construction*, p. 5; ADM 1/8370, *Instructions for Designing & Construction Departments*, 1 Jan. 1903.

6. ADM 1/7737, 3 Dec. 1904 (Feb. 1903); Hansard, 4, cxxxii, cols. 553-80 (23 March 1904); ADM 1/7661, 20 Oct. 1903.

7. See the anecdote in Esther Meynell, *A Woman Talking* (London, 1940), pp. 69-71.

8. NMM, WH1/44, Henderson to ----, 8 Nov. 1904, fo. 1.

9. *Ibid.*, fos. 1, 5.

10. ADM 1/7801, 5 Jan. 1905, *Rept. of the Naval Establishments Enquiry Committee*, p. 1, draft of letter to Treasury, p. 12.

11. ADM 1/8220, 12 July 1911, *Rept. on the Royal Corps of Naval Constructors*, p. 85.

12. H.C. (1906), lxx. 476-777; NMM, WH1/44, Henderson to ----, 8 Nov. 1904, fo. 6.

13. ADM 1/7807, 5 Jan 1905, *Rept., Naval Establishments Enquiry*, p. 3.

14. This last extended the practice since 1873 of handing structural materials directly over to the department heads.

15. H.C. (1906), lxx. 476-78; ADM 1/7807, *Rept., Naval Establishments Enquiry*, pp. 3-4; ADM 1/7990, 14 Jan. 1908, *3rd Rept. of the Subcommittee on Accounts and Staff*, pp. 7-8.

These measures were part of a larger reorganization in itself unrelated to the dockyards. The port commanders supervised the victualling, ordinance and coaling department, but this was found to be incompatible with their fleet duties and so was shifted to the superintendents, who consequently had to be given compensatory relief.

16. ADM 1/7807, *Rept., Naval Establishments Enquiry*, pp. 3-4, 11; H.C. (1906), lxx. 476.

17. *Ibid.*, 477; ADM 1/7807, *Rept., Naval Establishments Enquiry*, pp. 1-2; ADM 1/8881 (1906), pp. 9-10.

18. ADM 1/7737, 31 Dec. 1904 (Feb. 1903).

19. This includes one ship that was a very long time building (over three years). See Parkes, pp. 425-545 *passim*.

20. Marder, *Dreadnought to Scapa Flow*, I, pp. 36-40; ADM 1/7813, 14 Nov. 1905, "Navy and Dockyards," pp. 1, 2, 9; ADM 1/7814, 25 Nov. 1905; ADM 116/1026, 12 Jan. 1909; H.C. (1906), lxx. 78, 451-52; ADM 1/7731, 23 Jan. 1904.

21. ADM 1/7807, 5 Jan. 1905, *Preliminary Rept., Naval Establishments Enquiry Committee*, pp. 8, 1-15; ADM 116/1028, Marshall's minute, 19 Oct. 1906; H.C. (1906), lxx. 78.

22. ADM 1/7807, *Rept., Naval Establishments Enquiry*, pp. 4-5; Parkes, pp. 425-545 *passim*.

Forty-six battleships were built between 1893 and 1905, eighteen of them in private yards. H.C. (1905), xlviii. 69-72.

23. *Rept., Select Committee on Finance* [1913], p. v, *Evidence*, qq. 1166-1170; ADM 1/8046, 22 April 1909 (McLaren to McKenna, 21 April 1909, and covering letter, 22 April 1909).

Earle's, the Hull shipbuilders with whom Reed and Elgar were associated in the 1870s, failed in 1900 after making heavy losses on Admiralty contracts, and the Thames Ironworks in 1912 failed for the same reason after completing the dreadnought *Thunderer*.

24. D.K. Brown, *A Century of Naval Construction. The History of the Royal Corps of Naval Constructors 1883-1983* (London, 1983), p. 27.

25. ADM 1/7597, 2 May 1902; ADM 1/7674, 12 March 1903; Pollard and Robertson, pp. 12-23.

26. ADM 1/7993, 22 July 1908, *Rept. of the Committee on the Writing Staff* [1909], pp. 2-22. See also ADM 116/319; ADM 1/7659, 29 June 1903.

The British iron and steel industry may have been no less backward in adopting modern office equipment. See P.L. Payne, "The Emergence of the Large-Scale Company in Great Britain, 1870-1914," *Econ. Hist. Rev.*, 2, XX (1967), pp. 534-35.

27. ADM 1/7807, *Rept., Naval Establishments Enquiry*, pp. 1-6; H.C. (1906), lxx. 478.

28. ADM 1/7807, *Rept. of the Subcommittee on Stores*, pp. 5-6; H.C. (1906), lxx. 478.

29. ADM 1/7990, 14 Jan. 1908, *3rd Rept., Subcommittee on Accounts and Staff*, pp. 14-20; ADM 1/7993, 22 July 1908 (Sept. 1909, *Rept., Dockyard Writing Staff*, pp. 22-26); ADM 1/8377 (1913-14).

30. Watts, an Admiralty constructor, succeeded White at Elswick in 1885 and was brought back to the Admiralty when White retired in Jan. 1902. Selborne said that White's "jealousy of any rival" had prevented the rise of an "eminent assistant...who could take his place." ADM 1/7528, 27 Nov. 1901 (draft letter, Feb. 1901).

31. Marshall became Director of Dockyards when Sir James Williamson retired in 1905.

32. ADM 1/8220, *Rept., Royal Corps*, pp. 8-9, 11-14, *Evidence*, p. 51.

33. *Rept., Royal Corps*, pp. 5-6, 13-14, *Evidence*, p. 77.

34. *Rept., Royal Corps*, pp. 6-9.

35. Haas, "Best Investment," pp. 330-31.

36. The chairman and managing director of Fairfields, Alexander Gracie, was on the Inchcape Committee.

37. *Rept., Royal Corps*, p. 5, *Evidence*, pp. 27-29.

38. ADM 1/8272, 23 March 1912 (12 June 1912, *Rept. of the Committee on*

the Controller's Department, p. 1); ADM 1/8220, *Rept., Royal Corps*, p. 6.

39. The chairman was Sir George Murray, until 1911 Permanent Secretary of the Treasury and now a director of Armstrong-Whitworth, the giant munitions firm. Lord Inchcape was on the Committee also.

40. ADM 1/8272, 23 March 1912 (Churchill, minute of 1 Jan. 1912, p. 5-7).

41. ADM 1/8272, 23 March 1912 (12 June 1912, *Rept., Controller's Department*, p.4).

42. Contracts, formerly scattered among various offices in the Controller's department and elsewhere, were consolidated in the contract and purchase department (now transferred from the Parliamentary Secretary to the Additional Civil Lord). The consolidation had been recommended as long ago as 1887. *Repts., Dockyard Expenditure* [1887], p. 339.

43. ADM 1/8272, 23 March 1912 (12 June 1912, *Rept., Controller's Department*, p. 5.

44. ADM 1/8274, 7 Sept. 1912, pp. 6-8.

45. ADM 1/8272, 23 March 1912 (12 June 1912, *Rept., Controller's Department*, p. 5.)

There was a good deal of concern, however, about "the extent to which...[department heads] ought to be entrusted with devolved financial power." Safeguards were sufficient in the case of large repairs (approved by the Financial Secretary) but not in the case of alterations and additions. Instructions for the latter, it was decided in July 1914, were to be issued only upon an estimate and approval by the Permanent Secretary, superintending lords, and Financial Secretary. ADM 1/8333, 17 Oct. 1913, *passim*.

46. ADM 116/3453, Jackson to Balfour, 25 Nov. 1915.

47. Fisher was brought back as First Sea Lord when anti-German sentiment forced Prince Louis of Battenberg to resign in 1914.

48. Marder, *Dreadnought to Scapa Flow*, II, pp. 93-96.

49. He was elevated to the peerage as Baron Southborough.

50. The Ministry of Shipping until this time controlled mercantile shipbuilding.

51. For a fuller account see Haas, "Trouble at the Workplace," pp. 219-24.

52. Mortimer, p. 160; Hansard, 4, cxxxii, cols. 553-55, 558 (23 March 1904); H.C. (1906), lxx. 78; ADM 106/1028, Treasury to Admiralty, 27 Nov. 1906; Clegg, I, pp. 435-37.

53. Good, pp. 610-11; ADM 116/1028, 19 Oct. 1906; ADM 1/8332, 18 Aug. 1913.

54. ADM 116/1028 (Marshall's minute, 19 Oct. 1906; Treasury to Admiralty, 12 Oct. 1909); ADM 116/1218 (25 April 1913, pp. 5-6; 5 Aug. 1913).

55. Haas, "Trouble at the Workplace," pp. 221-22.

Bibliography

1. Manuscript Materials

a. Public Record Office

ADM 1.	Secretary's Department, In-letters.
ADM 2.	Secretary's Department, Out-letters.
ADM 3.	Admiralty Board Minutes.
ADM 7.	Miscellanea.
ADM 49.	Miscellanea.
ADM 83.	Controller, In-letters.
ADM 106.	Navy Board, Out-letters and Miscellanea.
ADM 116.	Miscellanea.
IND.	Secretary's Department, Digests.

b. British Library

Add. MSS. 38217, 38344.	Liverpool Papers.
Add. MSS. 40450, 40556.	Peel Papers.
Add. MSS. 40839.	Instructions for Commissioners of the Navy.
Add. MSS. 41368, 41374, 41433, 41434.	Martin Papers.
King's MSS. 44.	Survey of Dockyards, 1774.

c. National Maritime Museum

POR F.	Commissioner Hughes's letters to the Navy Board from Portsmouth yard.
ADM B.	Admiralty in-letter books from the Navy Board.
SAN.	Lord Sandwich's appointment books.
WH.	Henderson Papers.

d. Ministry of Defence, Naval Historical Library

8th Rept. of the Commission of Naval Revision.
N. Macleod, Extracts of Portsmouth Dockyard Records.

e. William Clements Library, University of Michigan.

Shelburne MSS., vols. 146, 151.

2. British Official Publications

a. Sessional Papers of the House of Commons

Further Proceedings Respecting the Committee on Accounts.	1801. v.
6th Rept. of the Commission of Naval Enquiry.	1803-04. iii.
9th Rept. of the Commission of Naval Enquiry.	1805. ii.
1st Rept. of the Commission of Naval Revision.	1805. ii.
2nd Rept. of the Commission of Naval Revision.	1806. v.
3rd Rept. of the Commission of Naval Revision.	1806. v.
3rd Rept. of the Commission on Fees.	1806. vii.
5th Rept. of the Commission on Fees.	1806. vii.
6th Rept. of the Commission on Fees.	1806. vii.
4th Rept. of the Commission of Naval Revision.	1809. vi.
5th Rept. of the Commission of Naval Revision.	1809. vi.
6th Rept. from the S.C. on Finance.	1817. iv.
8th Rept. from the S.C. on Finance.	1818. iii.
S.C. on Poor Rate Returns.	1825. iv.
3rd Rept. from the S.C. on Public Income and Expenditure.	1828. v.
Orders respecting Dockyard Establishments.	1831-31. xxvii.
Circular to Dockyards.	1847. xxxvi.
S.C. on Navy, Army, and Ordinance Estimates.	1847-48. xxi(1).
S.C. on the Navigation Laws.	1847-48. xx(2).
S.C. on the Dockyard Appointments.	1852-53. xxv.
Return of Appointments.	1856. xli.
Committee on Dockyard Economy.	1859(2). xviii.
Observations on the Report of the Committee on Dockyard Economy.	1860. xlii.
Royal Commission on the Control and Management of Naval Yards.	1861. xxvi.
Return of Yard and Factory Instructions.	1864. xxxviii.
Report from the Accountant General of the Navy.	1865. xxxv(1).

Report on Records of Accounts of Expenditure.	1865. xxxv(1).
S.C. on Admiralty Monies and Accounts.	1867-68. vi.
S.C. on the Board of Admiralty.	1871. vii.
Royal Commission on H.M.S. "Megaera".	1872. xv.
Admiral Sir Spencer Robinson, Letter to the First Lord.	1872. xxxix.
Committee on the Store Department.	1877. xxii.
Committee on the Professional Officers.	1883. xvii.
Committee on Contracts.	1884. xiv.
Return showing action of the Admiralty on the Ravensworth Committee Report.	1884-85. xlviii.
Committees on Admiralty and Dockyard Administration and Expenditure.	1886. xiii.
Committees on Admiralty and Dockyard Administration and Expenditure.	1887. xvi.
5th Rept. of the Committee of Public Accounts.	1890. x.
3rd Rept. of the Committee of Public Accounts.	1890-91. xi.
Royal Commission on Labour.	1893-94. xxxii.
Return of Naval Establishments.	1896. liv.
Return of Ships.	1905. xlviii.
Memorandum on recent and forthcoming changes.	1906. lxx.
Return of Shipbuilding in government and private yards.	1906. lxx.
Memorandum on Works in Progress.	1906. lxx.
Statement of Admiralty Policy.	1906. lxx.
Return of Building Costs.	1910. lxi.
S.C. on Estimates.	1913. vi.
Return of Building Costs.	1914. liii.
Epitome of the Reports from the Committee of Public Accounts.	1937-38. xiii.

b. Journals of the House of Commons

11th Rept. of the Commissioners on Crown Lands.	xlvii (1792).

c. Reports from Committees of the House of Commons

17th Rept. from the Select Committee on Finance.	xiii (1797).
31st Rept. from the Select Committee on Finance.	xiii (1797).

d. Other Official Publications

Hansard, *Parliamentary Debates.*
Rept. by the Comptroller and Auditor General: Ministry of Defence: Fleet Maintenance. (HMSO, 1990).

Statutes at Large.

3. Original Printed Materials

Barnes, G.R. and Owen, J.H. eds., *The Private Papers of John, Earl of Sandwich* (4 vols., Navy Records Society, 1932-38).

Bentham, Samuel, *Answers to the Comptroller's Objections on the Subject of His Majesty's Dockyards* (London, 1800).

Bentham, Samuel, *Desiderata in a Naval Arsenal* (London, 1814).

Bentham, Samuel, *Naval Papers and Documents* (London, 1827).

Bentham, Samuel, *Services rendered in the Civil Department of the Navy* (London, 1813).

Brassey, Thomas, *The British Navy: Its strengths, resources, and administration*, Vol. IV (London, 1883).

Brassey, Thomas, *Recent Naval Administration* (2nd edn., London, 1872).

Corbett, Julian S. and Richmond, H.W., eds., *Private Papers of George, second Earl Spencer* (2 vols., Navy Records Society, 1913-1914).

Hamilton, Richard Vesey, ed., *Letters and Papers of Admiral of the Fleet Sir Thomas Byam Martin* (3 vols., Navy Records Society, 1898-1903).

Laughton, John Knox, ed., *Letters and Papers of Charles, Lord Barham* (3 vols., Navy Records Society, 1907-11).

Markham, Clements, ed., *Selections from the Correspondence of Admiral J. Markham* (Navy Records Society, 1904).

Merriman, R.D., ed., *The Sergison Papers* (Navy Records Society, 1950).

Merriman, R.D., ed., *Queen Anne's Navy. Documents Concerning the Administration of the Navy of Queen Anne 1702-1714* (Navy Records Society, 1961).

A Scheme of Prices for Performing Shipwrights' Work (London, 1857).

A Scheme of Prices for Performing Smiths' and Millwrights' Work (London, 1857).

Smith, David Bonner, ed., *Letters and Paper of Admiral of the Fleet the Earl of St. Vincent whilst First Lord of the Admiralty 1801-1804* (2 vols., Navy Records Society, 1922-27).

4. Contemporary Pamphlets

Anon., *An Apology for English Ship-Builders* (London, 1833).

Anon., *An Explanation of the Conduct of Government in Instituting the School of Naval Architecture* (London, 1828).

Anon., *Review of the Course Pursued by the Shipbuilding Department of the Admiralty; between the Years 1832 and 1847* (Plymouth, 1847).

S[hrubsole]., W[illiam]., *A plea in Favour of the Shipwrights belonging to the*

Royal Dockyards Rochester, 1770).
[Somerset, Duke of and Grey, Frederick], *The Naval Expenditure from 1860 to 1861, and Its Results* (2nd edn., London, 1867).

5. Periodical Publications

Gentleman's Magazine.
Wall Street Journal.
World's Work.

6. Books and Articles

Abbott, A., *Education for Industry and Commerce in England* (Oxford, 1933).
Abell, Westcott, *The Shipwright's Trade* (Reprinted, Jamaica, NY, 1962).
Aldcroft, Derek H. and Richardson, Harry W., *The British Economy 1870-1939* (London, 1969).
Argles, Michael, *South Kensington to Robbins. An Account of English Technical and Scientific Education since 1851* (London, 1964).
Arthur, Charles B., *The Remaking of the English Navy by Admiral St. Vincent — Key to Victory over Napoleon* (New York, 1986).
Ashton, T.S., *An Economic History of England: The Eighteenth Century* (London, 1955).
Ashworth, William, *An Economic History of England 1870-1939* (London, 1960).
Ashworth, William, "Economic Aspects of Late Victorian Naval Administration," *Econ. Hist. Rev.*, 2, XXII (1969), 491-505.
Banbury, Philip, *Shipbuilders of the Thames and Medway* (Newton Abbot, 1971).
Barnaby, Nathaniel, *Naval Development of the Century* (London, 1904).
Barry, P., *Dockyard Economy and Naval Power* (London, 1863).
Bates, L.F., *Sir Alfred Ewing* (London, 1946).
Baugh, Daniel A., *British Naval Administration in the Age of Walpole* (Princeton, NJ, 1965).
Baxter, James Phinney, *The Introduction of the Ironclad Warship* (Cambridge, MA, 1933).
Bendix, Reinhard, *Work and Authority in Industry* (New York, 1963).
Bentham, M.S., *The Life of Brigadier-General Sir Samuel Bentham* (London, 1862).
Bienefeld, M.A., *Working Hours in Industry: An economic history* (London, 1972).
Binney, J.E.D., *British Public Finance and Administration 1774-92* (Oxford, 1968).
Bowley A.L. and Wood, George H., "*The* Statistics of wages in the United

Kingdom *during the* Nineteenth Century (Part XIV). Engineering *and* Shipbuilding," *Royal Statistical Soc. Journ.* (1906, 148-96.

Bowley, A.L., *Wages and Income in the United Kingdon since 1860* (Cambridge, 1937).

Breihan, John R., "The Addington Party and the navy in British Politics 1801-1806," in Craig. L. Symonds, ed., *New Aspects of Naval History* (Annapolis, MD, 1981), 163-89.

Brewer, John, *The Sinews of Power. War, Money and the English State, 1688-1783* (Cambridge, MA, 1990).

Brown, D.K., *A Century of Naval Construction. The History of the Royal Corps of Naval Constructors 1883-1983* (London, 1983).

Bruce, Alexander Balmain, *The Life of William Denny* (London, 1889).

Casey, Neil, "An Early organizational hegemony: methods of control in a Victorian naval dockyard," *Social Science Information*, 23 (1984), 677-700.

Childers, Spencer, *The Life and Correspondence of the Right Hon. Hugh C.E. Childers 1827-1896* (2 vols., London, 1901).

Chubb, Basil, *The Control of Public Expenditure* (Oxford, 1952).

Clapham, J.H., *An Economic History of Modern Britain* (3 vols., Cambridge, 1926-38).

Clegg, H.A. and others, *A History of British Trade Unions since 1889*, Vol. I, *1889-1910* (Oxford, 1964).

Coad, Jonathon G., *The Royal Dockyards 1690-1850. Architecture and Engineering Works of the Sailing Navy* (Aldershot, 1989).

Cotgrove, Stephen, *Technical Education and Social Change* (London, 1958).

Crimmin, P.K., "Admiralty Relations with the Treasury, 1783-1806: The Preparation of Naval Estimates and the Beginnings of Treasury Control," *Mariner's Mirror*, 53 (1967), 63-72.

Davies, C.S.L., "The Administration of the Royal Navy under Henry VIII; the Origins of the Navy Board," *Eng. Hist. Review*, LXXX (1965), 268-88.

Deane, Phyllis and Cole, W.H., *British Economic Growth 1688-1959* (Cambridge, 1962).

Dougan, David, *The History of North East Shipbuilding* (London, 1968).

Dunlop, John T. and Diatchenko, Vasilii P., eds., *Labor Productivity* (New York, 1964).

Ehrman, John, *The Navy in the War of William III, 1689-1697* (Cambridge, 1962).

Fabricant, Soloman, *A Primer on Productivity* (New York, 1969).

Fairbairn, William, *Treatise on Iron Shipbuilding; Its History and Progress* (London, 1865).

Garner, S. Paul, *Evolution of Cost Accounting to 1925* (University, AL, 1954).

Gilboy, Elizabeth W., *Wages in the Eighteenth Century* (Cambridge, 1962).

Haas, J.M., "The Best Investment Ever Made: The Royal Dockyard Schools, Technical Education, and the British Shipbuilding Industry, 1800-1914,"

Mariner's Mirror, 76 (1990), 325-35.

Haas, J.M., The Introduction of Task Work into the Royal Dockyards, 1775," *Journ. Br. Studies*, VII (1969), 44-68.

Haas, J.M., "Methods of Wage Payment in the Royal Dockyard 1775 -1865," *Maritime History*, V (1977), 99-115.

Haas, James M., The Pursuit of Political Success in Eighteenth Century England: Sandwich, 1740-71," *Bull. Inst. Hist. Res.*, XLIII (1970), 56-77.

Haas, James M., "The Royal Dockyards: The Earliest Visitations and Reform 1749-1778," *Hist. Journ.*, XII (1970), 191-215.

Haas, J.M., "Trouble at the Workplace: Industrial Relations in the Royal Dockyards, 1889-1914," *Bull. Inst. Hist. Res.*, LVIII (1985), 210-25.

Haas, J.M., "Work and Authority in the Royal Dockyards from the Seventeenth Century to 1870," *Proc. Amer. Phil. Soc.*, 124 (1980), 419-28.

Habakkuk, H.J., *American and British Technology in the Nineteenth Century. The Search for Labour-Saving Inventions* (Cambridge, 1962).

Hamilton, George, *Parliamentary Reminiscences and Reflections 1886-1906* (London, 1922).

Harris, J., *Unemployment and Politics; A Study of English Policy, 1886-1914* (Oxford, 1972).

Harris, William Snow, *Our Dockyards. Past and Present State of Naval Construction* (Plymouth, 1863).

Hogg, E.G., *The Royal Arsenal* (London, 1963).

Hovgaard, William, *Modern History of Warships* (reprint, London, 1971).

James, William, *The Naval History of Great Britain, from the Declaration of war by France in 1793, to the accession of George IV* (6 vols., London, 1878).

Jenks, Leland H., "Early Phases of the Management Movement," *Admin. Science Quart.*, V (1960), 421-47.

Jones, Leslie, *Shipbuilding in Britain (Cardiff, 1957)*.

Knight, R.J.B., "Sandwich, Middleton and Dockyard Appointments," *Mariner's Mirror*, 57 (1971), 175-92.

Lacour-Gayet, G., *La Marine Militaire de la France sous Le Regne de Louis XV* (Paris, 1910).

Lavery, Brian, *The Ship of the Line* (2 vols., London, 1987).

Lewis, Michael, *The Navy of Britain. A Historical Portrait* (London, 1948).

Littleton, A.C., *Accounting Evolution to 1900* (2nd edn., New York, 1966).

Littleton, A.C. and Yamey, B.S., *Studies in the History of Accounting* (Homewood, IL, 1956).

Lloyd, Christopher, *Mr. Barrow of the Admiralty. A Life of Sir John Barrow 1764-1848* (London, 1970).

Lyon, Hugh, "The relations between the Admiralty and private industry in the development of warships," in Bryan Ranft, ed., *Technical Change and British Naval Policy 1860-1939* (London, (1977), 37-64.

MacDougall, Phillip, *Royal Dockyards* (Newton Abbot, 1982).

Mackay, Ruddock F., *Admiral Hawke* (Oxford, 1987).

Mackesy, Piers, *The War for America, 1775-1783* (London, 1964).

Mallett, Bernard, *British Budgets, 1877-78 to 1912-23* (London, 1913).

Manning, Frederic, *The Life of Sir William White* (London, 1923).

Marder, Arthur J., *The Anatomy of British Sea Power* (reprinted, Hamden, CT, 1964).

Marder, Arthur J., *From the Dreadnought to Scapa Flow: The Royal Navy in the Fisher Era, 1904-1919*, Vol. I, *The Road to War* (London, 1961).

Marder, Arthur J., *Portrait of an Admiral: Life and Papers of Sir Herbert Richmond* (London, 1952).

McKendrick, Neil, "Josiah Wedgwood and Cost Accounting in the Industrial Revolution," *Econ. Hist. Rev.,* 2, XII (1970), 45-67.

Meynell, Esther, *A Woman Talking* (London, 1940).

Middleton, Richard, "Naval Administration in the Age of Pitt and Anson, 1755-1763," in Jeremy Black and Philip Woodfine, eds., *The British Navy and he Use of Naval Power in the Eighteenth Century* (Atlantic   Highlands, NJ, 1989), 109-27.

Miller, Francis H., *The Origin and Constitution of the Admiralty and the Navy Boards* (London, 1884).

Mitchell B.R. and Deane, Phyllis, *Abstract of British Historical Statistics* (Cambridge, 1962).

Morriss, Roger, *The Royal Dockyards during the Revolutionary and Napoleonic Wars* (Leicester, 1983).

Morriss, Roger, "Samuel Bentham and the Management of the Royal Dockyards, 1796-1807," *Bull. Inst. Hist. Res.,* LIII (1981), 226-40.

Mortimer, J.E.,   *History of the Boilermakers' Society, Vol. I, 1834-1906* (London, 1973).

Murray, Oswyn A.R., "The Admiralty," *Mariner's Mirror,* 23 (1937), 13-35, 129-47, 316-31; 24 (1938), 100-04, 204-25, 329-52, 458-78.

Musson, A.E., *The Growth of British Industry* (New York 1978).

Musson, A.E. and Robinson, Eric, *Science and Technology in the Industrial Revolution* (Manchester, 1969).

Oppenheim, M., *A History of the Administration of the Royal Navy and of Merchant Shipping in Relation to the Navy* (1896, reprinted, Camden, CT, 1961).

Parkes, Oscar, *British Battleships, "Warrior" to "Vanguard"* (London, 1956).

Parkinson, J.R., *The Economics of Shipbuilding in the United Kingdom* (Cambridge, 1960).

Parris, Henry, *Constitutional Bureaucracy. The Development of British Central Administration since the Eighteenth Century* (London, 1969).

Paullin, Charles Oscar, *Paullin's History of Naval Administration 1775-1914* (Annapolis, MD, 1968).

Payne, P.L., "The Emergence of the Large-Scale Company in Great

Britain, 1870-1914," *Econ. Hist. Rev.*, 2, XX (1967), 519-42.

Phillips, I. Lloyd, "Lord Barham at the Admiralty, 1805-06," *Maritime Hist.*, V (1977), 217-33.

Pollard, Sidney, "British and World Shipbuilding, 1890-1914: A Study of Comparative Costs," *Jour. Econ. Hist.*, XVII (1957), 426-44.

Pollard S., "The Decline of Shipbuilding on the Thames," *Econ. Hist. Rev.*, 2, III (1950), 72-89.

Pollard, Sidney, *The Genesis of Modern Management* (London, 1968).

Pollard, Sidney and Robertson, Paul, *The British Shipbuilding Industry, 1890-1914* (Cambridge, MA, 1979).

Pritchard, James, *Louis XV's Navy 1748-1762. A Study of Organization and Administration* (Kingston and Montreal, 1987).

Prothero, I.J., *Artisans and Politics in Early Nineteenth-Century London* (Folkestone, 1979).

Robertson, D.J., *The Economics of Wages and Distribution of Income* (London, 1961).

Robertson, Paul L., "Technical Change in the British Shipbuilding and Marine Engineering Industries, 1863-1914," *Econ. Hist. Rev.*, 2, XXVII, 222-35.

Rodger, N.A.M., *The Admiralty* (Lavenham, 1979).

Rodger, N.A.M., "The Dark Ages of the Admiralty, 1869-1885. Part I, 'Business Methods', 1869-74," *Mariner's Mirror*, 61 (1975), 328-50.

Rodger, N.A.M., *The Wooden World: An Anatomy of the Georgian Navy* (Annapolis, MD, 1986).

Roseveare, Henry, *The Treasury* (London, 1969).

Rowntree, B. Seebohm, *Poverty: A Study of Town Life* (2nd edn., London, 1902).

Sandler, Stanley, *The Emergence of the Modern Capital Ship* (Newark, DE, 1979).

Schaffer, B.F., "The Idea of the Ministerial Department: Bentham, Mill, and Bagehot," *Austral. Journ. Pols. and Hist.*, III (1957-58), 60-78.

Schloss, W.F., *Methods of Industrial Remuneration* (London, 1891).

Smith, Edgar C., *A Short History of Naval and Marine Engineering* (Cambridge, 1937).

Solomons, David, ed., *Studies in Cost Analysis* (Homewood, IL, 1968).

Sumida, Jon Tetsuro, *In Defence of Naval Supremacy. Finance, Technology and British Naval Policy, 1889-1914* (Boston, MA, 1989).

Tennyson D'Eyncourt, Eustace H.W., *A Shipbuilder's Yarn. The Record of a Naval Constructor* (London, 1948).

Thompson, E.P., *The Making of the English Working Class* (New York, 1963).

Tillett, Anthony and others, eds., *Management Thinkers* (Harmondsworth, 1970).

Ure, Andrew, *The Philosophy of Manufacturers* (London, 1835).

Urwick, L., *The Golden Book of Management* (London, 1956).

Urwick, L. and Breck, E.F.L., *The Making of Scientific Management* (3 vols.,

London, 1957).

Ward, J.T., *Sir James Graham* (London, 1967).

Webb, Paul, "Construction, repair and maintenance of the battle fleet of the Royal Navy, 1793-1815," in Jeremy Black and Philip Woodfine, eds. *The British Navy and the Use of Naval Power in the Eighteenth Century* (Atlantic Highlands, NJ, 1989), 207-19.

Wright, Maurice, *Treasury Control of the Civil Service 1854-1874* (Oxford, 1969).

7. Unpublished Dissertations

Knight, Roger Beckett, "The Royal Dockyards in England at the Time of the American War of Independence," Ph.D. diss., University of London, 1972.

Usher, Roland Green, "The Civil Administration of the British Navy during the American Revolution," Ph.D. diss., University of Michigan, 1942.

Williams, M.J., "The Naval Administration of the Fourth Earl of Sandwich," D. Phil. diss., Oxford University, 1962.

Index